2/7/91

Oxford Physics Series

General Editors

E. J. BURGE   D. J. E. INGRAM   J. A. D. MATTHEW

D0076208

Oxford Physics Series

The Solid State

An introduction to the physics of
solids for students of physics,
materials science, and engineering

THIRD EDITION

H. M. ROSENBERG
Fellow of St Catherine's College, Oxford

OXFORD UNIVERSITY PRESS

Oxford University Press, Walton Street, Oxford OX2 6DP

Oxford New York Toronto
Delhi Bombay Calcutta Madras Karachi
Petaling Jaya Singapore Hong Kong Tokyo
Nairobi Dar es Salaam Cape Town
Melbourne Auckland

and associated companies in
Berlin Ibadan

Oxford is a trade mark of Oxford University Press

Published in the United States
by Oxford University Press, New York

© *Oxford University Press, 1975, 1978, 1988*

First printed 1975
Second edition 1978
Third edition 1988
Reprinted (with corrections) 1989

British Library Cataloguing in Publication Data

Rosenberg, Harold Max
The solid state. — 3rd ed. —
(Oxford physics series, no. 9)
1. Solid state physics
I. Title
530.4'1   QC176
ISBN 0–19–851871–4
ISBN 0–19–851870–6 Pbk

Library of Congress Cataloging in Publication Data
Data available

Printed in Great Britain by
J. W. Arrowsmith Ltd, Bristol

Editor's foreword

THE extension of our understanding of the properties of solids at a microscopic level is one of the important achievements of physics in this century, and a study of solid state physics is now an essential ingredient of all undergraduate courses in physics, materials science, metallurgy, electrical engineering, and related topics. Dr. Rosenberg's book in this field succeeds in capturing the spirit of the stimulating lecture course that Oxford undergraduates have enjoyed for many years. It may be used as a text for a first course in solid state physics or as a follow up to a qualitative survey of the nature of condensed matter. Dr. Rosenberg's book requires only a fairly basic background in mechanics, electricity and magnetism, and atomic physics along with relatively intuitive ideas in quantum physics, but from such a modest platform of knowledge he is able to develop a remarkably mature perspective of the solid state. All the usual aspects are considered, but dislocations receive a particularly lucid treatment. By emphasizing the simple case and by avoiding complicated mathematical derivation, Dr. Rosenberg gives insights which many much larger and more expensive books in this area fail to provide.

J.A.D.M.

Preface

YET another book on solid state physics requires some explanation and excuse. Most solid state texts are fat and encyclopaedic. My aim was to write a thin volume and the result, whilst inevitably thicker than I had anticipated, is still reasonably slim. The treatment is intended for second-year undergraduates of physics, materials science, and engineering and it is an attempt to describe the basic elements of the physics of solids with as much continuity as is possible. Particular emphasis has been placed in the early chapters on the properties of waves in a periodic structure since this is the only basis on which to build a proper appreciation of diffraction, lattice vibrations, and electrons in metals and semiconductors.

Since the book is meant as an introductory text complicated proofs and points of detail have been omitted, although certain interesting topics which are not discussed or are not developed in the chapters are treated in the problems.

In this third edition the author has included a chapter on the properties of amorphous solids. Progress in this field in the past decade has made this one of the most exciting areas of solid state studies. No longer can a book on solids deal only with the properties of crystalline materials.

Many textual changes have been made to amplify or clarify certain points and short sections have been added on low-dimensional semiconducting structures and on magnetic materials.

Extra problems have been provided and some of these involve the use of a computer or a programmable calculator. These now enable students to tackle problems of physical significance which previously they would not have found possible.

A book like this can lay little claim to originality. I owe a great debt to the standard texts by Dekker, Kittel, and Wert and Thomson, as well as to several of the other books quoted in the bibliography.

I am very grateful to the many people who have suggested improvements, but particular thanks are due to Professor L. M. Slifkin who read through the second edition and made many valuable suggestions.

I should also like to thank many of my colleagues for reading and criticizing certain chapters. In particular these include S. L. Altmann, J. W. Christian, A. H. Cooke, G. Garton, W. Hayes, R. W. Hill, J. W. Hodby, F. N. H. Robinson, A. C. Rose-Innes, J. Singleton, and C. E. Webb.

I am very grateful to all those who kindly supplied me with photographs for the figures, to Mrs. B. Wanklyn and to Dr. D. Double who provided specimens for microphotography, and to Mr. A. G. King who gave valuable help in taking many of the micrographs.

Acknowledgement is also made to the Delegates of the Oxford University Press for permission to use some questions set in examinations of the University of Oxford.

Oxford   H. M. R.
June 1987

Contents

1. Atoms in crystals

NEARLY all the materials which we use today are crystals—collections of atoms or molecules assembled in some characteristic regular pattern. For an easy life physicists have nearly always limited their work to explanations of the properties of simple regular patterns of atoms—who can blame them—for even these can be extremely complex; so they have been concerned with the behaviour of a piece of rocksalt, or a diamond, or perhaps a single crystal of silicon or sodium. It is with the basic properties of such systems that we too must begin. But like life, most materials are neither regular nor simple. A piece of metal is rarely a single crystal, it is usually an agglomerate of small crystallites packed together. The lattice structure of such small crystallites, or even of large crystals, is rarely perfect. There will be regions where atoms are wrongly placed, or perhaps missing; impurities, accidental as well as intentional, will be present. In some cases the regularity of atomic arrangement extends over such a short distance that the material can hardly be considered to have any crystalline structure—it is a glass. (We deal with this very important class of materials in Chapter 15.) Finally in the higher stages of complication and confusion we have very important structural materials of a composite nature such as wood and fibre-glass. Clearly a full understanding of their properties is a formidable problem.

It must be admitted at the outset that the only materials whose properties we can understand and explain in depth are single crystals of high purity. On the basis of this understanding we must then use our judgment and what knowledge we have to see how best we can account for the properties of more complicated structures.

1.1. Atoms in crystals

The regular arrangement of atoms in a material is, of course, the distinguishing feature of a crystal as compared with an amorphous substance. However, many scientists, particularly physicists, tend to disregard the details of the atomic arrangement and they often represent a small volume of a crystal as a simple cubic array, as shown in Fig. 1.1. Some even seem to regard the actual details of the atomic arrangement as being rather *earthy*, and perhaps best left to metallurgists and engineers who can cope with such irrelevant complexities.

Although only one element (polonium) is known to crystallize in the simple cubic form this structure does have the virtue that it is easy to draw; but

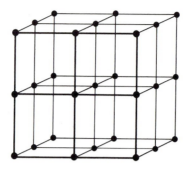

FIG. 1.1. The simple cubic structure.

as we shall see, many of the important properties of solids can be explained only if details of the structure of atoms in crystals are taken into account. We shall therefore give a brief description of the crystal structures of some of the more common simple materials. In a stable configuration of atoms the interatomic separation is usually between 0·2 nm and 0·3 nm (2–3 Å). The precise arrangement depends largely on the form of the electron clouds which surround the central positive nucleus of every atom. In many cases the crystal structure of a solid changes at certain temperatures and the examples which we give below are those which are stable at room temperature.

1.2. Cubic structures

Atoms do not arrange themselves on a simple cubic pattern because this is a very open structure. There are many empty spaces, and as a rule atoms tend to pack together more closely. If we examine the simple cubic arrangement to see how it might accommodate more atoms, the most obvious empty space is that at the centre of the cube. If this is filled we have the body-centred-cubic (b.c.c.) structure (Fig. 1.2(a)). Note that it can be considered as being two interlinked simple cubic lattices which are displaced with respect to one another along the body-diagonal of one cube by half the length of the diagonal (Fig. 1.2(b)). Many metals have a b.c.c. structure. The most important is iron, but in addition there are Cr, Mo, W, and also the alkali metals Li, K, Na, Rb, and Cs.

Certain simple binary compounds have what is essentially the b.c.c. arrangement in which the centre of the cube is occupied by one type of atoms and the corner sites are occupied by the others. This is usually called the caesium chloride structure.

We should here differentiate between the terms 'structure' and 'lattice' which are often used synonymously. By 'structure' we mean the actual pattern of the arrangement of atoms in space. The term 'lattice', although loosely used to denote the atoms in a crystal, should be reserved to describe that

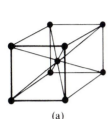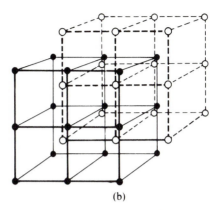

(a)　　　　　　　　　　　　　　　　　　(b)

FIG. 1.2. (a) The body-centred cubic structure. (b) The body-centred cubic structure may be considered to be two interlinked simple cubic arrays.

network of points which show the simple translation vectors on which the structure is based. Thus Fe (b.c.c. structure) also has a b.c.c. *lattice* since a translation of an atom from the corner of the cube to the various points on that lattice is necessary in order to describe the atomic arrangement. The CsCl structure described above, however, has as its basis a simple cubic *lattice*, composed of pairs of Cs and Cl ions. It does *not* have a b.c.c. lattice because a translation from the corner of a cube to the body centre would not result in the correct structure since the atoms at those two sites are different.†

1.3. Close-packed structures

If we consider the atoms as hard spheres then the most efficient packing in one plane is the close-packed arrangement shown in Fig. 1.3. There are two simple ways in which such planes can be layered on top of one another to form a three-dimensional structure. In the layer which is illustrated the atoms are all at positions A and a second layer could nest above the A layer by being either in the spaces marked B or in those marked C. Let us suppose that the second layer is in the B spaces. Where can the succeeding layers be placed? There are two possible arrangements each of which is a very important type of structure.

1.4. Hexagonal close-packed structure

The simplest arrangement is for the third layer to be laid down in positions A, i.e. immediately above the atoms in the first layer, and the fourth to be at

---

† It might be thought by this argument that Fe also has a simple cubic lattice composed of pairs of Fe ions. However, since the ions are now identical the full symmetry of the structure would not be covered by this description.

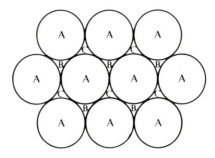

FIG. 1.3. The close-packed array of spheres. Note the three different possible positions, A, B, and C for the successive layers.

B and so on. This gives a layering of the type ABABAB... (or, of course, we could have ACAC...). This arrangement, in which the pattern is repeated at every alternate layer, is called the hexagonal close-packed (h.c.p.) structure. The hexagonal form of the arrangement is clearly seen from Fig. 1.4. Note that the basal planes of the hexagon (i.e. the planes parallel to the page in Fig. 1.4(a)) are unique. There are no other planes in other directions which are equivalent, i.e. which have the same close-packed pattern. The *hexagonal axis* is normal to the basal plane. Many metals have the h.c.p. structure. Among them are Mg, Zn, Cd, Ti, and Ni.

1.5. Face-centred cubic structure

There is another way in which successive close-packed planes of the type shown in Fig. 1.3 can be stacked. The first and second layers can be in positions A and B as before, but the third layer, instead of reverting to the A positions as in the h.c.p. arrangement, can be placed above the C positions. The fourth layer is then put in the A positions which are immediately above the atoms in the lowest plane, and so the pattern continues as ABCABC... with the pattern repeating at every third layer (Fig. 1.5(a)).

This type of array is called the face-centred cubic (f.c.c.) structure. At first sight there seems to be nothing cubic about it. This is because the close-packed layers which we have been discussing do not correspond to the ordinary faces of a cube. A face-centred cubic unit cell drawn with conventional cube axes as the basis is shown in Fig. 1.5(b). As its name implies the structure is the same as the ordinary simple cubic lattice (Fig. 1.1) with the addition of an extra atom at the centre of each cube face. The close-packed layers which we discussed in the previous paragraph are the body diagonal planes of the cube and one of these is shown in Fig. 1.5(c). The f.c.c. structure has a very important property which the h.c.p. arrangement does not possess—there are four equivalent close-packed planes as can be seen from a study of Fig. 1.5(c).

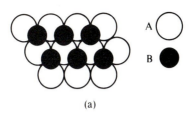

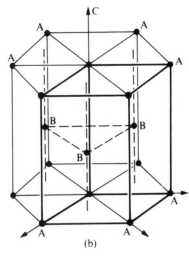

FIG. 1.4. Hexagonal close-packing. (a) The layering sequence ABABA... (b) The three-dimensional arrangement of the atoms which shows the hexagonal pattern more clearly.

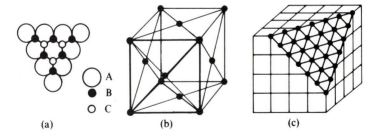

FIG. 1.5. The face-centred cubic structure. (a) The layering sequence ABCABCA... (b) The unit cell based on a cube. (c) The close-packed planes of (a) are the body-diagonal planes of the cube. (After D. Hull, *Introduction to Dislocations*, Pergamon Press, Oxford.)

It should nevertheless be noted that although in the h.c.p. structure the basal plane is the only one which is close-packed, the actual fraction of space filled in h.c.p. is the same as in f.c.c.

Face-centred cubic structures are typical of many metallic elements, e.g. Cu, Ag, Au, Al, Pd, and Pt.

1.6. The diamond structure

The crystal structure of diamond can be derived from the f.c.c. lattice although it is not itself a close-packed arrangement. Formally it may be described as being built up from two interpenetrating f.c.c. lattices which are offset with respect to one another along the body-diagonal of the cube by one quarter of its length (Fig. 1.6(a)). Whilst this definition accurately describes the position of the atoms, it is not very useful when one tries to envisage the positions of the atoms relative to one another. A more useful model is to consider each atom to be at the centre of a tetrahedron with its four nearest neighbours at the four corners of that tetrahedron (Fig. 1.6(b)). From Fig. 1.6(a) it can be seen that if the 'central' atom of the tetrahedron is on one f.c.c. sub-lattice then the four 'corner' atoms are all on the other sub-lattice. It should also be noted that since all the atoms are equivalent *each* of them can be thought of as being the centre atom of a tetrahedron, although it would require a rather extended diagram to show this.

The tetrahedral arrangement in the diamond crystal in which each atom has four equally-spaced nearest-neighbours is a consequence of the special type of electron sharing (covalent bonding) which occurs in these structures. This is described in a later section of this chapter.

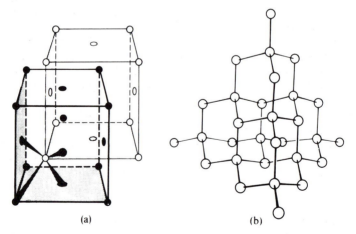

(a)   (b)

FIG. 1.6. The diamond lattice showing (a) how it is formed from two interpenetrating f.c.c. lattices, (b) the tetrahedral arrangement of the atoms.

Apart from diamond, the other elements which have the same structure are also tetravalent—germanium and silicon. Tin also has a diamond-structure allotrope 'grey tin' which is stable below about $-40\,°C$.

If the two interpenetrating lattices are of two different elements the atoms on different sub-lattices are of course no longer equivalent although they will still yield a similar tetrahedral arrangement. This is then called the zincblende (ZnS) structure and it is typical of the 3–5 semiconducting crystals such as InSb, GaAs, and GaP.

The structures which we have discussed in the preceding sections are the simplest types which occur, although of course a whole hierarchy of structures of increasing complexity may be found, particularly in biological materials. But even though many materials do not crystallize precisely in the form of the simple lattices which have been described, some are quite close to them. For example, some materials are nearly cubic, but the sides of the 'cube' are not all equal, or in other cases the sides may be equal but the angles of the 'cube' are not exactly right angles. It should also be emphasized, as has already been remarked, that some materials change their crystal structure at certain temperatures. For instance, from among the examples we have given, iron becomes f.c.c. and nickel becomes h.c.p. at high temperatures, whereas at very low temperatures (about 30 K) sodium transforms to a complex mixed structure.

1.7. The unit cell

An entire single crystal is produced by a process of repeating the simple atomic patterns such as those which we have just described. Each of these elementary building blocks (e.g. the body-centred cube in Fig. 1.2(a)) is called a unit cell.

It is often possible to break down the unit cells into smaller simpler arrangements which contain fewer atoms (often, in elements, a single atom based on a single lattice point) and which, when repeated, will also make up the crystal structure. These are called primitive cells. The choice of primitive cell or unit cell is purely a matter of geometrical convenience. Thus the crystal can be considered to be based on a lattice, at each point of which is a primitive cell. This cell is sometimes called a *basis*.

Since unit cells often contain more than one atom, the size of the unit cell is usually larger than the nearest-neighbour interatomic spacing in a crystal and care must be taken not to confuse the unit cell dimensions with the atomic spacing.

1.8. The nomenclature for directions and for planes in crystals

Although a single crystal is formed by a regularly repeated pattern of atoms, the material can still exhibit anisotropy, i.e. certain properties, such as the electrical conductivity or the velocity of sound might have different values along different crystallographic *directions*. Or again, if we deform the crystal, say, in tension, then this deformation proceeds as certain preferred *planes* of atoms slide over one another. In order to discuss the properties of crystals we therefore need a simple way of labelling unambiguously the directions and planes within a crystal. On first reading the systems might seem to be unduly pedantic but they have the advantage that they can be used, not merely for the simple types of structure which we have been discussing, but also for more complicated arrangements and in particular for lattices which do not have cubic symmetry. We shall, however, only give examples for crystals with cubic symmetry. The nomenclature we shall describe is also particularly well adapted to be used in the mathematical theory of the diffraction of waves in crystals.

1.9. Designation of directions

Let the sides of the unit cell have lengths and directions given by the vectors a, b, and c. In the cubic system these will of course be mutually perpendicular to one another and equal in length, but in the general case this will not be so. We wish to label a certain direction in the crystal lattice. This is indicated in Fig. 1.7 by the direction of the vector z which passes through the origin, O,

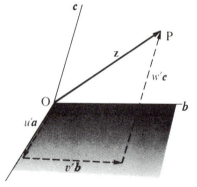

FIG. 1.7. The designation of directions. The direction of the vector z is given in terms of its projections on the unit-cell axes as described in the text.

of the axes of the unit cell. We proceed by starting at O and we go a distance $u'a$ along the a axis, $v'b$ parallel to the b axis and $w'c$ parallel to the c axis so that with a suitable choice of u', v', and w' we finally end up somewhere along the vector z, say at P. $u'a$, $v'b$, and $w'c$ are, of course, the projections of OP on to the three axes. The three numbers u', v', and w' are then reduced† to the set

† By dividing each of them by their highest common factor.

of simplest integers u, v, w, and the direction of z is labelled as $[uvw]$ using square brackets. If a negative direction was used to reach z then this is indicated by a bar over the appropriate letter, e.g. $\bar{v}$. The complete set of equivalent directions in a crystal is denoted, using angular brackets, by $\langle uvw \rangle$. In the cubic system u, v, and w are proportional to the direction cosines of the vector z

As examples, the cube edge direction a in Fig. 1.7 would be denoted by $[100]$, that of direction b by $[010]$ and the negative direction of a would be $[\bar{1}00]$. The general set of all the cube edge directions would be $\langle 100 \rangle$. The face-diagonals of the cube are $\langle 110 \rangle$ and the body-diagonals are $\langle 111 \rangle$. These are the only important directions which need be remembered.

1.10. Designation of atomic planes—Miller indices

The planes are labelled in a similar way to the directions. Let the plane in which we are interested cut the three axes of the unit cell (Fig. 1.8) at $u'a$,

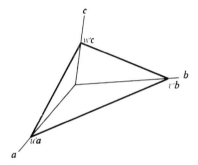

FIG. 1.8. Miller indices. The shaded plane is described in terms of the reciprocals of its intercepts with the unit-cell axes as described in the text.

$v'b$, and $w'c$ respectively. We then reduce the *reciprocals*, $1/u'$, $1/v'$, $1/w'$ to the simplest integers h, k, l, and the plane is labelled as (hkl), using round brackets. These are called the Miller indices of the plane. Once again a bar is used for a negative direction. The general set of planes of a particular type is placed within curly brackets $\{hkl\}$.

In the cubic system the Miller indices of a plane are just the reciprocals of the intercepts of that plane on each of the three axes, reduced to the simplest integers. For example, the set of all cube faces is $\{100\}$; the planes which pass through the face-diagonals of two parallel sides are $\{110\}$ and those which pass through three face diagonals on adjacent sides are $\{111\}$. These are shown in Fig. 1.9.

The Miller index system enables the spacings between adjacent planes to be

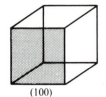
(100)

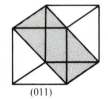
(011)

(111)

FIG. 1.9. The most important planes in a cubic system.

calculated very easily. For example, in the cubic system with a unit cell dimension of a, the spacing d between the $\{hkl\}$ planes is

$$d = a/(h^2 + k^2 + l^2)^{\frac{1}{2}}.$$

The angle ϕ between two planes (or directions) in a cubic system with indices $h_1k_1l_1$ and $h_2k_2l_2$ is given by

$$\cos\phi = \frac{h_1h_2 + k_1k_2 + l_1l_2}{(h_1^2 + k_1^2 + l_1^2)^{\frac{1}{2}}(h_2^2 + k_2^2 + l_2^2)^{\frac{1}{2}}}$$

The most important planes are those with low index numbers and these are the ones which are the most widely spaced.

Note also that in the cubic system the $[hkl]$ vector is perpendicular to the (hkl) plane.

For the hexagonal crystal lattice four indices are used $(hkil)$. These are derived by using the intercepts on four axes a,b,c,d, three in the basal plane at 120° to each other, and the fourth one normal to the basal plane (Fig. 1.10).

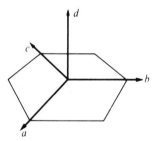

FIG. 1.10. Miller indices for the hexagonal lattice. The intercepts of the plane on three basal axes a,b,c and the one hexagonal axis d are used to derive the indices.

The reciprocals of these intercepts are then reduced to simple integers as before, and the third index is chosen to be minus the sum of the first two. Although, of course, one of these indices must be redundant, the four-figure system is a more convenient one to use for the hexagonal lattice. Thus the basal plane is (0001) and $\{10\bar{1}0\}$ designates the set of hexagonal faces. An analogous system is used for designating directions.

1.11. Binding forces in crystals

What holds the atoms together so that they form a crystal? There must be an attractive force between atoms, because otherwise solids would not be formed; but there must also be a repulsive force which prevents the atoms from getting too close together. In general terms the stability of the crystal lattice depends on a balance between this repulsive force, which is dominant when the atoms are very close, and the attractive force which remains important when they are slightly farther apart.

The stable position for the atoms is most satisfactorily understood by considering how the potential energy of a pair of atoms varies with their separation. The repulsive force gives rise to a positive potential energy since under its influence work has to be done on the system in order to bring the atoms closer together. We shall assume that this energy varies as some inverse power of the atomic separation, r, say as A/r^n, although it is sometimes written in the form $A \exp(-r/\alpha)$. The attractive force gives a negative potential energy of the form $-B/r^m$. This tends to zero when the atoms are far apart and it increases negatively as they approach one another. Curves for these two contributions to the potential energy, as a function of the interatomic spacing are shown in Fig. 1.11, together with that for the total potential energy $A/r^n - B/r^m$. It can be seen that, as drawn, the total potential energy passes through a minimum. The atomic separation r_0 at which this minimum potential energy occurs is, of course, the stable spacing for the pair of atoms. It is at this separation that the negative and positive forces annul one another. It should be noted that a minimum in the potential energy (and hence a position of *stable* equilibrium) can only arise if the power n in the repulsive term is greater than m in the attractive term. If $m > n$ an unstable situation would develop and all the atoms would collapse together.

1.12. The dissociation energy

The zero of energy (which is of course quite arbitrary) used in Fig. 1.11 is taken to be the energy which the two atoms have when they are far apart. The depth of the energy minimum at the equilibrium spacing is a measure of the work which must be done to remove an atom, originally stably bound in the lattice, right away from the crystal. This is called the dissociation energy or the cohesive energy and it may be determined experimentally by calorimetric measurements.

1.13. The repulsive force

Let us now consider the physical mechanisms which give rise to the forces between atoms.

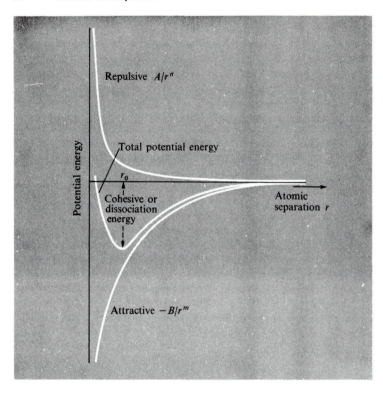

FIG. 1.11. The potential energy of two atoms as a function of their separation r. There are both positive (repulsive) and negative contributions. These yield a minimum in the potential energy at a separation r_0, which should correspond to the equilibrium separation. The value of the potential energy at r_0 is the dissociation energy.

The repulsive force is most quickly dealt with. When atoms are brought very close together the electron clouds around neighbouring atoms are made to overlap. There is a mutual repulsion between like charges and so the electron cloud around one atom repels that around its neighbour. The closer the atoms and their charge clouds approach one another, the stronger does this repulsive force become.† Materials in which the atoms have a large electron radius (e.g. potassium) tend to have a large atomic separation.

---

† This is an oversimplification because this is Chapter 1! We should also take account of the fact that as the electron clouds start to overlap, the electron energies are shifted. Some electrons will therefore need to increase their energy. Hence there will be a repulsion—see section 8.19, p. 140.

1.14. The attractive force

There are several mechanisms which can give rise to an attractive force between atoms and in many materials one of these is dominant and it can be used to characterize the type of the atomic interaction. In other substances no one attractive mechanism can be singled out and so a combination of them must be considered.

The electron charge distribution around the atomic nucleus is subdivided into various clouds or 'shells', each of which surrounds the nucleus and contains a fixed number of electrons (further discussion is given in sections 7.2 and 11.4). Most binding mechanisms rely on the fact that these various charge clouds are more stable if they contain a certain special number of electrons ('closed shells'). To achieve this, the outer partially filled shells can either lose or gain electrons fairly easily. The details of the binding process depend to a large extent on the ease with which these outer electrons can be removed from or attracted to the atom to form closed shells.

1.15. Ionic binding

This is the most straightforward type of binding to understand—and it is the only one which can be computed with any confidence. A well-behaved binary ionic crystal (e.g. NaCl) consists of two sets of atoms whose inner, filled electron shells are surrounded by a partially filled shell. In one set the outer shell contains only one electron (e.g. Na); it is said to be strongly electropositive because very little energy is required to remove the outer electrons and this leaves a positively charged ion, e.g. Na^+. The other set of atoms has an outer shell which is nearly filled, apart from one empty state (e.g. Cl). This can attract an electron to fill the shell, thereby becoming an electronegative ion Cl^-. This transfer of an electron† from one type of atom to the other, thereby enabling both of them to become ions with closed electron shells, is the essential feature of the ionic solid (Fig. 1.12(a)). Throughout the crystal there is an alternating pattern of positive and negative ions, such as $Na^+Cl^-Na^+Cl^-$.‡ The attraction between the ions may then be accounted for in a straightforward way as being due to the electrostatic interaction between unlike electric charges. It should be noted that whilst energy is actually *required* to form $Na^+ + Cl^-$ the energy of the system is reduced because of the electrostatic interaction. If the ionic separation is r, then the electrostatic energy will be $(ve)^2/(4\pi\varepsilon_0 r)$ where v is the valency of the electro-positive ion, and the attractive force will be $(ve)^2/(4\pi\varepsilon_0 r^2)$. A simple calculation shows that

---

† We have described the simplest situation in which the outer electron shells either lack or have just *one* electron. Divalent and trivalent ionic crystals are composed of atoms whose outer shell lacks or has two or three electrons.

‡ Although this ionic arrangement is simple cubic the anion–cation *pairs* form a f.c.c. *lattice*.

for an ionic separation of 0·3 nm the attractive energy is of the order of 5×10^{-19} J (a few electronvolts) and this is typical of the interaction energies between atoms in crystals.

Of course a proper calculation of the binding energy of an ionic crystal should include not merely an expression for the interaction between a pair of adjacent ions, but it should also take account of the attractive and repulsive effects of the more distant ions as well. This is a quite straightforward, although tedious, extension of the calculation and it is usually taken into account by multiplying the interaction term for neighbouring ions by a geometrical correction factor called the Madelung constant, which depends on the details of the crystal structure. For the NaCl lattice the Madelung constant is almost exactly 1·75 and for other structures it tends to lie between 1·6 and 1·8.

1.16. Covalent bonding

In ionic crystals, as we have seen, electrons are transferred from one atom (e.g. Na) to another (e.g. Cl) in order that both will thereby have closed shells of electrons. Another type of interaction arises if the electron-cloud geometry is distorted sufficiently so that electrons are continually being *shared* between neighbouring atoms, thereby forming completed shells on all of them. By sharing, we mean that the electron spends as much time near one atom as it does near another one—and so it is impossible to say that the electron now belongs to a *particular* atom. This type of sharing is especially important if each atom has a half-filled outer shell; e.g. the carbon atom has four outer electrons in a shell which can accommodate eight. We have already described (section 1.6) that in the diamond lattice each carbon atom may be considered to be at the centre of a tetrahedron with its nearest neighbours at the four corners of the tetrahedron. Each of these four outer atoms can be thought of as sharing one of its electrons with the central atom, (Fig. 1.12(b)) thereby making up a closed shell of eight electrons. It should be noted that the 'central' atom (which as has already been pointed out is not in any privileged position, but is actually equivalent to the 'corner' atoms) will share each of its four electrons with the four corner atoms, thereby contributing to their closed shells of eight electrons.

The calculation of covalent bonding energies and forces is very hazardous because it demands a computation of the amount of overlap of the electron charge clouds for neighbouring atoms—it involves what is called an exchange integral—and it is not possible to calculate this with any confidence. The fact that the amount of overlap, and hence the reduction in energy due to the overlap, varies very rapidly with interatomic spacing implies that the binding forces associated with covalent bonding are usually very strong ones. The crystals are usually hard and they have high melting points.

In diamond the covalent bonding is very effective. In other crystals, and especially in compounds, neighbouring atoms might not share their electrons

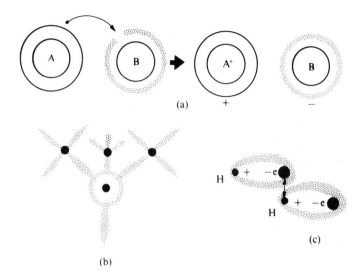

(a)

(b)

(c)

FIG. 1.12. Attractive binding mechanisms between atoms or ions. (a) Ionic binding; the outermost electron of one atom migrates to the neighbouring atom thereby producing two ions with completed outer shells. (b) Covalent bonding. Electrons are shared between atoms so that they all effectively have filled outer shells. (c) Hydrogen bonding. The electron from a hydrogen atom which is covalently bonded may be shared unequally with the other atom, so that the hydrogen has a net positive charge. It can then be attracted to a negatively charged atom in another molecule.

equally; i.e. an electron from atom A might spend more time near atom B, than an electron from B spends near A. If this occurs then atoms A and B will be partially ionized—and there will be electrostatic forces between them. The binding will therefore not be completely covalent but it will be *partially* ionic in character. This *mixed* binding is very prevalent in compounds which have the zincblende structure (section 1.6). We can envisage a whole spectrum of different crystals in which there is a gradation in the type of binding from being completely ionic (electron transfer, but no sharing), to mixed (unequal sharing) and finally to covalent bonds (equal sharing).

1.17. Hydrogen bonding

Hydrogen has only one electron and it can therefore only be covalently bonded to one other atom. If, however, this second atom is strongly electronegative (fluorine and oxygen are the most important examples) then the electron from the hydrogen spends most of its time on the other atom and the hydrogen is left positively charged (Fig. 1.12(c)). It can then be attracted to another electronegative atom on a neighbouring molecule. It is this attraction between molecules (or sometimes between parts of a large molecule) which is known as the hydrogen bond. The cohesive forces in ice are due to hydrogen bonding.

1.18. Metallic binding

In a metal the outer electrons of an atom are able to escape from that atom quite easily, but there are no electro-negative atoms present to attract and confine them as there are in an ionic crystal. When such atoms are brought together to form a metal crystal the charge clouds of neighbouring atoms overlap so that it is easy for an electron to migrate from one atom to the next. Since this can occur for the outer electrons on all the atoms there is no electron build-up around any particular atom. These migrating electrons are called conduction electrons. We can consider the metal as a fixed lattice of positive ions permeated by a smeared-out negative charge cloud of conduction electrons. It is not immediately obvious that such a system will have a lower energy than a set of individual atoms, each with its full complement of electrons, but since metals exist it would appear that the reduction in energy which occurs with a system of positive ions in a sea of conduction electrons more than compensates for the energy required to remove each of those electrons from its original atom. There will therefore be a tendency (i.e. a force of attraction) for the metal atoms to move sufficiently close together so that the reduction in energy of the system is a maximum. This binding is sometimes described (although not very usefully) as an attraction between the positive ions and the negative sea of electrons.

1.19. van der Waals or molecular binding

All the attractive mechanisms which we have so far discussed have relied in one way or another on electron transfer between one atom and its neighbours. This can occur only if the energy required to remove an electron from an atom or a molecule is not too high. If, however, the electrons in an atom or a molecule are all in closed shells they are very tightly bound and they cannot escape. This is the state of affairs in atoms of the inert gases, He, Ne, A, Kr, and Xe and also in some organic molecules. Yet under suitable conditions even these materials will solidify since there is one attractive mechanism which is always available. It arises because there is always some relative motion between the electrons around an atom or a molecule and the central, positively charged nucleus. If the electron cloud was static and it was spherically symmetrical with respect to the nucleus, then at a distance from the atom there would be no electric field because the effect of the nucleus would be just cancelled out by the electrons. But because the electron cloud is always vibrating with respect to its nucleus an exact cancellation does not occur. This means that the atom or molecule will have an electric dipole moment, which will be continuously changing as the electron charge cloud alters its shape. It can be shown that between a set of fluctuating dipoles there is always an attractive force called the van der Waals' force which tends to bind them together (see Problem 1.9). This force is usually rather weak, but it provides a mechanism for solidification in substances which have closed electron shells. Because the van der Waals force is small, such materials are characterized by being soft and by having low melting points.

PROBLEMS

1.1. The density of b.c.c. iron is 7900 kg m^{-3}, and its atomic weight is 56. Calculate the side of the cubic unit cell and the interatomic spacing.

1.2. The density of f.c.c. gold is 19 300 kg m^{-3} and its atomic weight is 197. Calculate the separation between the close-packed planes. On the assumption that the atoms may be thought of as spheres which just touch one another, estimate the atomic radius.

1.3. The interatomic spacing of silicon (diamond lattice) is 0·235 nm. Calculate the density (atomic weight of Si = 28).

1.4. Calculate the angle between the (111) and the ($\bar{1}\bar{1}0$) planes in a cubic lattice. If the unit cell dimension is 0·3 nm, what is the spacing between parallel planes of each of these types?

1.5. Show that the (hkl) plane is perpendicular to the [hkl] direction in a cubic lattice.

1.6. The cohesive energy of a sodium chloride crystal is 6 eV per Na$^+$Cl$^-$ ion pair. The ionization energy of Na is 5 eV and the electron affinity of Cl is 3·75 eV. Neglecting the energy of repulsion estimate the interionic spacing. (The electron affinity of Cl is the binding energy of the last electron of Cl$^-$.)

1.7. The potential energy of a pair of atoms in a crystal is of the form $A/r^9 - B/r$, when their separation is r. The equilibrium separation is 0·28 nm and the dissociation energy is 8×10^{-19} J. Calculate A and B. Find the effective modulus of elasticity for the pair of atoms and calculate the force which would be necessary in order to reduce their spacing by 5 per cent.

1.8. The lattice energy, E, per ion pair of LiF may be written in the form $E = A \exp(-r/\alpha) - Me^2/(4\pi\varepsilon_0 r)$, where M is the Madelung constant, e is the electronic charge, r is the separation of neighbouring ions and A and α are constants. Calculate E for LiF. ($M = 1·75$, $\alpha = 0·02$ nm; lattice parameter of LiF = 0·4 nm.)

1.9. The van der Waals' binding is due to the mutual interaction of fluctuating atomic electric dipoles. The electric field at a distance r from a dipole of moment p is of the order of $p/\varepsilon_0 r^3$. If the polarizability (eqn 13.2) of an atom at r is α, show that the mutual dipolar energy of the two atoms is of the order of $-\alpha p^2/\varepsilon_0^2 r^6$.

2. Waves in crystals

2.1. Introduction

As we have seen in the previous chapter, a crystal can be considered to be a periodic structure built up by the regular repetition of a particular unit cell. To a very large extent solid state physics is concerned with the manner in which waves are propagated through such periodic structures. Under what conditions can the waves travel through the crystal without being scattered? Or, if there is scattering, in what direction will it occur? The waves may be of several different types—they may be X-rays, or matter waves associated with the electrons in the material, or they may be lattice vibrational waves (phonons). Whatever type of wave we consider, the general principles involved are the same.

2.2. The Bragg construction

Before we give a mild mathematical treatment of waves in crystals we present a partial solution because it is probably familiar to the reader. This is the Bragg construction which is used to determine the angles at which X-ray diffraction lines or spots should be observed. In this construction it is assumed (with no physical justification) that the X-rays which enter a crystal at an angle θ to a certain set of parallel planes are 'reflected' at successive planes as indicated in Fig. 2.1. For any arbitrary angle θ, however, the emergent rays which have

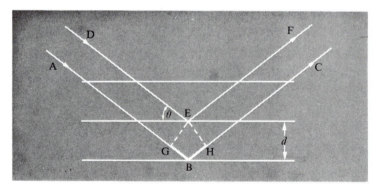

Fig. 2.1. The Bragg construction to determine the angles for diffraction. A plane wave is incident at an angle θ and is assumed to be 'reflected' at successive layers of atoms. Diffraction (i.e. constructive interference) occurs if after 'reflection' the rays from neighbouring planes remain in phase with one another. For the two rays shown this implies that $GB + BH$ must be a whole number of wavelengths.

been reflected by successive planes will not be in phase with one another and on average they will cancel out. In fact, there will only be a diffracted beam if the rays which have been 'reflected' by each plane emerge in phase with one another. Thus the extra path travelled by a particular ray ABC compared with that travelled by the ray DEF which has been reflected by the plane above must be a whole number of wavelengths. In Fig. 2.1 this extra path is equal to GB + BH. If the separation of the planes is d, then it is a simple matter to show that GB + BH = $2d \sin \theta$, and hence to observe a diffracted beam at an angle θ we must have

$$2d \sin \theta = n\lambda, \tag{2.1}$$

where n is an integer called the *order* of the diffraction and λ is the wavelength.

This is the famous Bragg formula and the angles θ for which it holds are called the Bragg angles. If eqn (2.1) cannot be satisfied for any type of plane in the crystal then diffraction cannot occur and the X-rays will pass through the crystal undeviated. It is particularly important to note that if λ is too long then (2.1) can never be satisfied since we must always have

$$n\lambda/2d \leqslant 1. \tag{2.2}$$

Thus λ must not be greater than twice an interplanar spacing, otherwise no diffraction can occur.

Whilst the Bragg formula is very useful it should be realized that it does not take full account of the physics of the problem. In particular, it is obvious that we cannot really assume that X-rays with wavelengths of the order of 0·1 nm can be *reflected* by planes which are composed of atoms whose spacing is of a similar order of magnitude. So far as the X-rays are concerned, these planes are *bumpy* and they cannot really be considered to be good reflecting surfaces.

2.3. Diffraction by a discrete lattice

To obtain more insight into the fundamentals of the problem we must treat the crystal as a lattice of discrete atoms. Each atom can interact with any incident wave. We *assume* (and it is only an assumption) that as each atom is excited by a wave, it re-radiates the wave in all directions in phase with the initial excitation at that particular atom (Fig. 2.2(a)). The amplitude of the beam in any direction is then obtained by summing the contributions in that direction from each atom, taking into account the phase differences between them.†

---

† This is, of course, an application of Huygens' principle of superposition which is used in optics. This is often presented as if it were an obvious truth—which it is not. A discussion is given by Lipson and Lipson, *Optical physics* (2nd edn), ˙p. 145 (Cambridge University Press, 1981).

In Fig. 2.2(a) this is illustrated for the simple case when the incident wave front is parallel to a line of atoms. The circles represent the positions of the

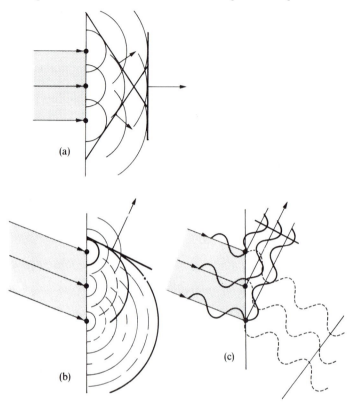

FIG. 2.2. Diffraction by a discrete lattice. The atoms are assumed to re-radiate in phase with the incident radiation. The arcs represent the position of the re-radiated wave crests at any instant and common tangents to the arcs give the direction of the wave fronts of the diffracted beams. (a) Normal incidence—two diffracted and one un-deviated beam are shown. (b) Oblique incidence—one diffracted beam is shown. (c) The situation in (b) is redrawn showing the actual re-radiated wave trains from each of the three atoms both in the direction of the diffracted beam (upper right) where they are all in phase and also in an arbitrary direction (lower right) for which there is destructive interference. For three atoms the destructive interference is not very satisfactory but for a much larger number of atoms it would be more complete and so the diffracted beam would be much sharper.

This figure only illustrates the case for a sheet of atoms perpendicular to the page when a diffracted beam will be observed for any incident angle. For a general two-dimensional or three-dimensional lattice diffraction only occurs for those angles of incidence which satisfy the Bragg formula.

wave crests at any instant and so along the tangents which are common to all circles the waves are all in phase. The tangents therefore indicate the positions of the wave fronts of the diffracted beams. In the figure three of these are shown. One corresponds to a ray which is transmitted without deviation and the others are the diffracted beams.

Fig. 2.2(b) and (c) show the more general case when the beam is *not* incident normally to the line of atoms. In Fig. 2.2(c) the actual wave forms are shown for this example—both in the direction of the diffracted beam, when they are all in phase—and also in an arbitrary direction where they are not in phase and where (if sufficient atoms were drawn) complete destructive interference would occur.

In a two- or three-dimensional lattice the directions of the diffracted beams will correspond to diffraction at the Bragg angles. At these angles the intensity of the diffracted beams will have a maximum value. It is clear, however, that if the direction of observation is *nearly*, but not quite, normal to the tangents of the circles drawn in Fig. 2.2(a), some radiation will still be detected, because the rays from individual atoms will not completely cancel one another; the diffracted beam therefore has a finite angular breadth, or spread. A finite breadth of the beam cannot be deduced from the Bragg construction since that treatment dispenses with the idea of discrete atoms.

2.4. Wave vectors

In general discussion a wave is usually characterized by its frequency and wavelength. However, in the mathematical treatment of waves it is much more convenient to use the *wave vector* k instead of the wavelength λ. We shall define k as $2\pi/\lambda$. It is a vector directed along the path of the wave propagation.†

k is more useful than λ because if we need to resolve the wave along two (or more) directions then it is the resolved part of k along those directions which must be used to find the components. λ cannot be resolved directly. Similarly, if two waves with vectors k_1 and k_2 combine, then the resultant wave vector is $k_1 + k_2$. Nevertheless, it is the wavelength which is more readily visualized rather than its reciprocal, and we shall tend to switch between one and the other in our discussion.

2.5. The mathematics of diffraction

As we have already discussed, the general principle which is used to calculate the diffraction of all types of waves in crystals is that the amplitude of the

† It should be noted that crystallographers usually define the numerical value of the wave vector as $1/\lambda$, so that the spatial part of the wave is represented by $A \sin 2\pi k \cdot r$ or in complex notation $A \exp(2\pi i k \cdot r)$. In reading the literature, however, there should be no confusion if it is remembered that the factor 2π must always occur, either explicitly or implicitly. Thus if a 2π is shown, then $k = 1/\lambda$; if it is not shown, then $k = 2\pi/\lambda$.

waves which are scattered (or re-emitted) by the individual atoms may be superimposed.

In a large perfect crystal the algebraic sum of these amplitudes (Fig. 2.2(c)) in any arbitrary direction will be zero, and only in very special directions (those where the Bragg conditions hold) will there be a non-zero amplitude, i.e. there will be a diffracted beam. If the conditions for diffraction are not satisfied, the beam will pass through the crystal without deviation. The larger the crystal the more sharply will any diffracted beam be defined, since the destructive interference between the atoms will be more efficient, even for small deviations from the Bragg angles.

If, however, there are any atomic irregularities in the crystal structure then there will be fewer atoms that can contribute to the constructive interference and so the intensity of the peak will be reduced, especially at higher angles of diffraction. The effect is similar to that produced by increasing the temperature (section 2.12).

In the following mathematical development it is convenient to use the complex form to express the amplitude of a wave at a point denoted by a vector r (with reference to some arbitrary origin O), as $A \exp\{i(k_0 . r - \omega t)\}$, where k_0 is the wave vector†, ω is the angular frequency of the radiation, and t is the time.

Consider a regular array of atoms (Fig. 2.3) situated at positions r' relative to O. They are excited, or illuminated, by a plane wave which will have an amplitude of $A \exp\{i(k_0 . r' - \omega t)\}$ at each value of r'. We assume, as has already been discussed, that the atoms re-radiate spherical waves in phase with the incident radiation. Although the wavelength of these secondary waves

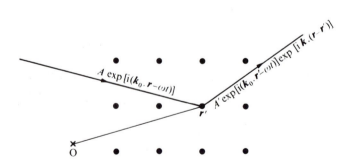

FIG. 2.3. Calculation of diffraction from a discrete lattice. An atom at a position given by the vector r' re-radiates an incident wave of wave vector k_0 as one with wave vector k.

† See footnote to section 2.4.

will be unchanged, their direction will in general be different from k_0—we shall denote it by k. We need to calculate the amplitude of this secondary wave at some point of observation r which we assume to be sufficiently distant from the atom that the section of the wave front we are considering may be treated as if it were plane. The vector from the point r to an atom at r' is $r - r'$. As it leaves the atom the wave has an amplitude which is proportional to

$$\exp\{i(k_0 . r' - \omega t)\}.$$

Hence the amplitude at the observation point is

$$A' \exp(ik_0 . r') \exp\{ik . (r - r')\} \exp(-i\omega t).$$

This can be rewritten as

$$A' \exp(ik . r) \exp\{i(k_0 - k) . r'\}, \tag{2.3}$$

where we have dropped the term in t which may here be set equal to zero without any loss of generality.

Provided that the observation point r is sufficiently far away from the crystal that the wave vectors k from all atoms to r may be assumed to be in the same direction, then the total effect at r due to all the atoms in the crystal is obtained by summing a set of terms as in (2.3) over all lattice points r'; i.e.

$$\text{total amplitude at } r \text{ of } k = A' \exp(ik . r) \sum_{\substack{\text{all atoms} \\ r'}} \exp\{i(k_0 - k) . r'\}.$$

This can be written in the form

$$\text{amplitude } (k) = A' \exp(ik . r) \int_{\text{all space}} f(r') \exp\{i(k_0 - k) . r'\} \, dr' \tag{2.4}$$

where $f(r')$ is a periodic function which has the value unity at a lattice site and which is zero elsewhere. (This assumes that the scattering occurs at point atoms which have no structure, see section 2.13 (p. 29).) The integral in (2.4) is called the Fourier transform of $f(r')$.

We now need to study the conditions which must be satisfied in order that (2.4) should be non-zero. These will then also be the conditions for diffraction to be observed.

2.6. Fourier transforms

There is nothing difficult about Fourier transforms and since they are so useful in dealing with any problem in which waves interact with a periodic structure it is worthwhile coming to terms with them as early as possible.

We shall not give a full mathematical treatment here. A more detailed discussion is given in Lipson and Lipson, *Optical physics* (2nd edn), Chapter 3 (Cambridge University Press, 1981).

The reader is probably familiar with the ideas of the Fourier analysis of a complicated wave shape. If $f(\alpha)$ is the amplitude of a complicated wave form with a period of 2π which extends to $\pm\infty$, f can be expressed as a sum

$$f(\alpha) = A_1 \sin\alpha + A_2 \sin 2\alpha + \cdots + A_n \sin n\alpha + \cdots + B_1 \cos\alpha + B_2 \cos 2\alpha + \cdots,$$

(2.5)

where the extra terms in 2α, 3α, etc. are harmonics of the fundamental. The coefficient A_n of the term in $n\alpha$, where n must be integral, can be found from

$$A_n = \frac{1}{\pi}\int_{-\pi}^{\pi} f(\alpha)\sin n\alpha\, d\alpha$$

(2.6)

and a similar expression is used to derive the coefficients B_n. The integral (2.6) is used because $\int_{-\pi}^{\pi} \sin m\alpha \sin n\alpha\, d\alpha$ is zero unless $m=n$ (and $\int_{-\pi}^{\pi} \sin m\alpha \cos n\alpha\, d\alpha = 0$ for any m, n). Hence if (2.5) is substituted into (2.6) the integral picks out the contribution which the term in $n\alpha$ makes to the function $f(\alpha)$.

Instead of f being periodic in the angle α, we can change to a one-dimensional spatial variable x so that $f(x)$ is periodic with a characteristic wavelength λ_c. This could be analysed into waves of the form $A_n \sin(2\pi nx/\lambda_c)$ or $A_n \sin(nk_c x)$, where k_c is the wave number† $2\pi/\lambda_c$. Then A_n may then be obtained in a similar way to (2.6) by substituting $\alpha = k_c x$

$$A_n = \frac{k_c}{\pi}\int_{-\pi/k_c}^{\pi/k_c} f(x)\sin(nk_c x)\, dx.$$

(2.7)

As before, A_n can only be non-zero if $f(x)$ contains a term in nk_c.

We can extend (2.7) to three dimensions and we also use complex notation in order to include both the sine and cosine terms in the most convenient manner. The integral in (2.7) then becomes

$$\int f(\mathbf{r})\exp(in\mathbf{k}_c\cdot\mathbf{r})\, d\mathbf{r}$$

or more generally,

$$\int f(\mathbf{r})\exp(i\mathbf{K}\cdot\mathbf{r})\, d\mathbf{r}.$$

‡(2.8)

---

† This is a wave *number* and not a wave *vector* because we are working in one dimension.

‡Although we shall not discuss the problem, (2.8) is also the Fourier transform even if $f(\mathbf{r})$ is not a periodic function over all space and, indeed, the term Fourier transform is often restricted to the case where f is not periodic or does not extend over all space.

The periodic function $f(r)$ which is our present concern is the atomic lattice. This has a fundamental wavelength λ_f $(=2\pi/k_f)$ which is equal to the unit-cell dimension. The integral (2.8) is therefore only non-zero if K is of the form nk_f; i.e. an integral number of waves of wavelength $2\pi/(nk_f)$ must fit into the length $2\pi/k_f$, the wavelength of the fundamental.

What do we mean by a wave 'fitting in' to the lattice? If we have a wave with a wave vector K which is travelling in some arbitrary direction in space then it can be resolved into three component waves travelling along three mutually perpendicular axes. The wave vectors of each of these three components are just the components of the original wave vector taken along the same three directions.

For a wave to 'fit in' to the lattice we need an integral number of wavelengths of each component to fill the corresponding dimension a_1, a_2, a_3, of the unit cell of the lattice (Fig. 2.4(a), (b)),

$$\text{i.e. } n_1\lambda_1 = a_1, \quad n_2\lambda_2 = a_2, \quad n_3\lambda_3 = a_3,$$

where the subscripts refer to the three axes. We can write this as

$$2\pi n_1/k_1 = a_1, \quad \text{or } k_1 = n_1(2\pi/a_1), \tag{2.9}$$

and similarly for k_2 and k_3, where k_1, k_2, and k_3 are the components of K.

2.7. The reciprocal lattice

The integral (2.8) is of the same form as the diffraction integral in (2.4) and we therefore see from (2.9) that we get constructive interference (i.e. diffraction) if K in (2.8) or $(k_0 - k)$ in (2.4) are such that their components, when resolved along the main axes of the crystal are an integer $\times 2\pi/$unit cell length.

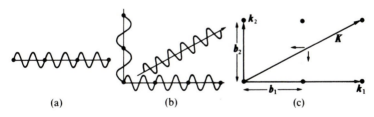

(a)   (b)   (c)

FIG. 2.4. Wave 'fitting in' to a lattice. An integral number of wavelengths must fill the unit cell. (a) In one dimension. (b) In two dimensions the wave *vector* must be resolved along the two appropriate directions and the waves corresponding to those two components must fit into the unit-cell dimensions in those two directions. (c) The same as (b) but shown directly in terms of the resolved parts k_1, k_2, of the original wave vector K. These must be an integer $\times$ the basic reciprocal lattice vectors b_1, b_2, respectively.

It is therefore useful to construct a network whose unit cell lengths are $2\pi/a_1$, $2\pi/a_2$, and $2\pi/a_3$. This is called the *reciprocal lattice*.† A wave vector whose length and direction are such that it can be drawn by connecting any two points of the reciprocal lattice will have components which are an integral number of corresponding unit cell lengths of the reciprocal lattice (Fig. 2.4(c)). This wave vector will therefore define a direction along which diffraction is permitted. The vectors which join two reciprocal lattice points are called *reciprocal lattice vectors*. The sides of the unit cell of the reciprocal lattice are called *basic vectors* of the reciprocal lattice.

If we again refer to the integral in the diffraction equation (2.4) we see that the wave vector in the exponent is $(k_0 - k)$. Thus diffraction occurs if the difference between the wave vectors of the incident and the scattered rays is equal to a reciprocal lattice vector. It should be particularly noted that the actual wave vectors themselves of either wave do *not* necessarily have to be reciprocal lattice vectors.

2.8. Geometrical construction for diffraction

In ordinary diffraction experiments the energy and hence the wavelength of the radiation is unchanged by diffraction, and so the absolute value (or modulus) of the wave vector is also unaffected. The directions of the possible incident and diffracted rays can therefore be found by a very simple construction. Draw a set of reciprocal lattice points (Fig. 2.5) and take a line joining any two of them, say AB. This is a reciprocal lattice vector and hence is a possible value of $(k_0 - k)$. Draw its perpendicular bisector CD. Now from A draw AE, equal in length to the modulus of the wave vector of the radiation, so that it cuts the bisector CD at E. Then if AE represents the direction of the incident beam, BE is the direction of the diffracted beam, because AB (a reciprocal lattice vector) is, of course, equal to the vector difference $AE - BE$. We can see from this construction that diffraction can only occur if k_0 is greater than half a basic reciprocal lattice vector. This is a generalization of the inequality (2.2).

The components (hkl) of the reciprocal lattice vector (in units of the basic vectors of the reciprocal lattice) are often used to designate a particular diffraction. Thus in the cubic system the (111) diffraction corresponds to a value of $(k_0 - k)$ equal to the body diagonal of the reciprocal lattice basic cube. The reader should verify that this also corresponds to Bragg 'reflection' from

---

† Note that this is a simplified definition which is only valid for cubic, orthorhombic, and tetragonal lattices. The general formulae for the lengths and directions b_1, b_2, b_3 of the unit cell of the reciprocal lattice are $b_1 = 2\pi a_2 \wedge a_3/(a_1 \wedge a_2 . a_3)$; $b_2 = 2\pi a_3 \wedge a_1/(a_1 \wedge a_2 . a_3)$; $b_3 = 2\pi a_1 \wedge a_2/(a_1 \wedge a_2 . a_3)$.
Crystallographers again usually omit the 2π.

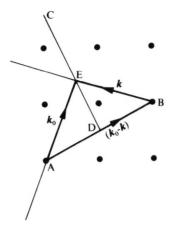

FIG. 2.5. The Laue construction for diffraction. The points represent the reciprocal lattice. The incident and diffracted beams have wave vectors k_0 and k respectively. Diffraction occurs whenever the vector $(k_0 - k)$ is equal to a reciprocal lattice vector—in this case **AB**.

the corresponding (111) planes of the crystal and, in general, the (hkl) diffraction is that from the (hkl) planes.

Some confusion, however, can arise for higher order diffractions. It will be recalled that the *order* of a diffraction is given by the integer n in the Bragg formula (2.1). It is the number of integral wavelengths for the path difference between rays diffracted from successive planes. A little consideration should show that if the first-order diffraction is given by a reciprocal lattice vector (hkl) then the nth order will correspond to $(nh\ nk\ nl)$. The diffraction planes in the lattice, however, are still (hkl). Unlike the Miller indices for the designation of atomic planes the diffraction indices are *not* reduced to the smallest integers, otherwise the important information of the order of the diffraction would be lost. Thus the *second*-order diffraction from the (111) planes will be designated (222).

It is clear from a study of Fig. 2.5 that for any arbitrary direction of the X-ray beam relative to the crystal, no diffraction will be observed. As the crystal is rotated, diffraction will occur for certain special orientations and these could be detected as a series of spots on a photographic emulsion. This is the basis of the rotating crystal method of diffraction.

2.9. The powder method

If a powdered specimen is used, instead of a single crystal, then there is no need to rotate the specimen, because there will always be some crystals at an orientation for which diffraction is permitted. The diffracted beam will

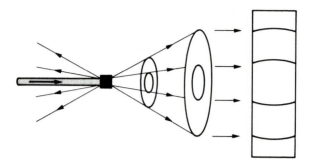

FIG. 2.6. The powder method of diffraction. A monochromatic beam of X-rays is incident on a powdered specimen and so there will always be some crystallites at an orientation which is suitable for diffraction. The diffracted beam therefore forms a series of cones. These will give a set of arcs on a strip of photographic film mounted around the specimen. Note that there are also diffracted beams in the backward direction.

therefore diverge from the specimen as a series of cones which will intersect a photographic film mounted around the sample as arcs of circles (Fig. 2.6). This is the *Debye–Scherrer* or *powder* method.

2.10. The Laue method

Instead of using a monochromatic beam of radiation we could use a broad continuous spectrum. If this is incident on a single crystal then there is no need for it to be rotated because, for any angle of incidence relative to the crystal axes, there will always be a suitable wavelength in the beam for which diffraction is permitted. This is the *Laue* method of diffraction, and it is the one which is commonly used to determine the orientation of a single crystal. It is not used in complex structural determinations because the continuous spectrum can give a complicated pattern of spots which would be very difficult to interpret.

In the preceding discussions we have assumed that the unit cells of the crystal are composed of atoms which are all identical and which are all exactly at their correct positions in the lattice. In general neither of these assumptions is justified and we must consider how the diffraction pattern will then be modified.

2.11. Broadening of lines by dislocations

First let us consider the effect of imperfections in the crystal lattice. There are always several types of geometrical defect present (see Chapter 3) however carefully a crystal is prepared. One of the most important of these is the dislocation, and its main property, so far as diffraction is concerned, is to divide up the crystal into a mosaic of crystallites which are slightly misoriented with respect to one another (section 4.24). If the crystal is set so that one part of it is exactly at the correct orientation for diffraction, another part will not be positioned correctly until it is turned through a small angle. Thus the diffraction pattern, instead of being very sharp, is broadened. The more disorder there is in the crystal, the greater will be the broadening. A detailed study of the breadth of X-ray patterns has been developed into a very useful technique for studying defects in crystals, particularly in severely cold-worked metals.

2.12. Effect of thermal vibrations

Even if we were able to obtain a 'perfect' crystal the atoms would not all be at their correct sites because, due to their thermal energy, they are vibrating It might at first be thought that the thermal vibrations would also lead to a broadening of the diffracted beam, but in fact this does *not* occur. The reason for this is as follows. At any instant a shift of the diffracted beam, because some atoms are displaced in a certain direction, will be cancelled by others which at that moment have moved in the opposite direction. Thus (to first order at least) the beam is *not* broadened. What is observed is that the *intensity* of the beam is diminished as the temperature is raised. This may be understood when one realizes that at higher temperatures there are fewer atoms at, or close to, their equilibrium sites on the crystal lattice at any instant, and hence the contribution to the diffracted beam will be reduced. The temperature dependence of the intensity is given by the *Debye–Waller factor*. If I_0 is the diffracted intensity from a perfect lattice the intensity of the beam when the mean square displacement of the atoms is $\langle u^2 \rangle$ is given by

$$I = I_0 \exp \left\{ -\tfrac{1}{3} \langle u^2 \rangle (k_0 - k)^2 \right\}. \tag{2.10}$$

A proof is given by Kittel (1986, p. 604).

As a material is cooled the atoms never become completely still. There always remains the zero-point motion demanded by quantum mechanics. Thus the X-ray intensity even at very low temperatures should never attain that of the perfect lattice. Indeed very careful experiments show that the intensity at low temperatures is not exactly as predicted by (2.10) if a value of $\langle u^2 \rangle$ is used which has been calculated from the classical thermal energy of the crystal. This is one of the most direct methods of demonstrating the existence of zero-point motion.

2.13. Effect of different types of atoms on the diffraction pattern

In this discussion we have tacitly assumed that the crystal lattice can be considered as a special arrangement of atoms or molecules each of which may be assumed to be a *point* scatterer (or re-radiator). This is, of course, an over-simplification. If we are dealing with X-rays we must take account of the fact that the atoms are surrounded by an electron cloud which actually interacts with the radiation. The details of this interaction will depend on the precise form of the electron cloud, i.e. on the particular type of atom we are considering. In general the more electrons there are in an atom the stronger will be the interaction and the diffracted beams will be more intense. Thus heavy atoms tend to give much stronger X-ray lines than light atoms and X-ray diffraction from atoms lighter than, say, carbon tends to be difficult. This can be a problem, particularly in the structure determination of organic substances. The scattering power of a particular type of atom is given by the *atomic scattering factor*. It is possible to estimate this from a knowledge of the electronic configuration of a particular atom, but the calculation is beyond the scope of this book.

2.14. Diffraction by crystals of molecules

A further complication in the X-ray diffraction pattern arises if the lattice sites are not occupied by single atoms, but by molecules–perhaps large complicated organic molecules such as those of a protein or DNA. The overall crystal structure can be fairly simple, but the detailed structure of the individual molecules at each lattice site might be extremely complex. The diffraction pattern which is obtained is then the overall lattice structure modulated (or multiplied) by the diffraction pattern of the molecular building block.† For example, let us assume that we have a crystal which, if it were composed of simple atoms, would give a diffraction pattern as in the lower part of Fig. 2.7(a). We now take an individual molecule which would give a pattern as in Fig. 2.7(b). The diffraction pattern from a crystal with the same lattice structure as (a) but made up from the molecules (b) rather than the single atoms, would then give a pattern as in Fig. 2.7(c). The diffraction pattern of the molecule is *sampled* at those places where a diffraction spot from the simple lattice should occur. The intensity of the diffraction spots now varies widely over the diffraction pattern. The important problem for crystallographers and molecular biophysicists is not so much to decide on the *overall* structure of the crystal, but to transform these *sampled* spots back to the molecular structure of the *unit*. In other words the intensities of the spots are used as a discontinuous set of readings taken at regular intervals over the diffraction pattern of the individual molecule. For biological material this is a complex task which still involves a considerable amount of trial, error, and intuition. It can only be done in a reasonable time with a computer.

2.15. Missing orders

Let us now return to our lattice of simple atoms again. It is found that if, for example, the diffraction pattern of various materials with cubic structures

---

† This is a consequence of the *convolution theorem* of Fourier transforms, see Lipson and Lipson, *Optical physics* (2nd edn), Chapter 3 (Cambridge University Press, 1981).

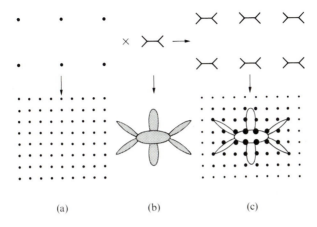

Fig. 2.7. Diffraction by crystals of complicated molecules. The crystal can be considered to be made from a point lattice (upper (a)) *convoluted* with the structure of the molecule (upper (b)). This gives the complete structure (upper (c)). The corresponding diffraction patterns of the three structures are shown on the lower parts of the figure. The pattern of the crystal (c) is the same as that of the point lattice (a) but its intensity has been modulated by the pattern of an individual molecule (b).

are compared then certain lines or spots which are present in some patterns are found to be absent in others. The particular missing lines depend on whether the crystals have a body-centred or a face-centred cubic structure. These missing lines are called missing or forbidden orders. The effect is due to the fact that along certain directions within the crystal extra atoms in the unit cell which are, say, midway between the atoms at the corners of the cell can re-radiate so that they are just out of phase with these corner atoms. There is therefore no diffracted beam in that direction, whereas in a simple cubic lattice one would be observed. For example, in Fig. 2.8(a) the body-centred atoms of the b.c.c. lattice form an extra set of atomic planes (Y) midway between the crystal planes formed by the atoms at the cube corners (X). Returning to the Bragg construction (Fig. 2.8(b)) we see that a diffracted beam which has a path difference of an odd number of wavelengths between planes X will be out of phase by an odd number of *half* wavelengths with the beam diffracted at the Y planes. The beams from the X planes therefore will interfere destructively with those from the Y planes. If however there was a path difference of an *even* number of wavelengths between the X planes (Fig. 2.8(c)) then there would be a path difference of an even number of half wavelengths between the X and Y planes and the beams would still reinforce one another so that diffraction would be observed. This is just one example

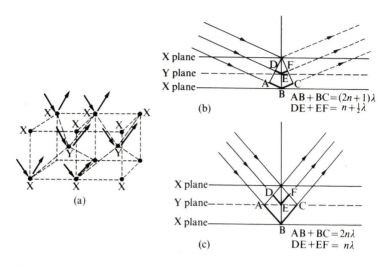

FIG. 2.8. Missing orders in the b.c.c. lattice. (a) The body-centred atoms Y form an extra set of Bragg planes midway between the planes of X atoms. (b) If the path difference AB + BC between neighbouring X planes is an *odd* number of wavelengths then the rays from the Y planes will be out of phase (DE + EF) with them and there will be no diffracted beam in that direction. If, however, the incident angle is changed as in (c) so that there is an *even* number of wavelengths between beams from neighbouring X planes then they will still be in phase with the beams from the Y planes and the order will be observed.

of the way in which forbidden orders arise. The effect is not limited to cubic lattices. Clearly the details of the forbidden orders will depend on the planes of atoms which are producing the diffracted beam.†

2.16. Diffraction of other kinds of waves; electron and neutron diffraction

Although our treatment of the diffraction of waves has been quite general we have so far concentrated on its application to the diffraction of X-rays in crystals. Similar considerations, however, will apply to any other types of waves which can interact with the atoms or ions. In particular we must consider the matter (or de Broglie) waves associated with electrons and with neutrons. Just as with X-rays, if the wave vector associated with the de Broglie wavelength is more than half the basic reciprocal lattice dimension (i.e. approximately 10^{10} m^{-1}) then it is possible to observe diffraction although of course the practical details are quite different for electron and for neutron diffraction.

† For the (hkl) diffraction there will be a forbidden order *unless*, for a b.c.c. lattice $h + k + l$ is even, or for a f.c.c. lattice, h, k, and l are *either* all even *or* all odd, counting zero as even.

The de Broglie wavelength λ of a particle is given by $\lambda = h/p$, where p is the momentum of the particle. Thus the wave vector $k = p/h$. In terms of the (non-relativistic) kinetic energy E, $k = (2mE)^{\frac{1}{2}}/h$, and so $E = h^2k^2/2m$. If k is of the order of 10^{10} m^{-1} then E will be about 5×10^{-19} J (3.8 eV) for electrons and about 3.3×10^{-22} J (2×10^{-3} eV) for neutrons.

Unfortunately, much higher energies have to be used if we wish to observe electron diffraction because the interaction of electrons with the atoms in a crystal is so strong that electrons with an energy of a few eV would be completely absorbed by a specimen. In order that an electron beam can penetrate even a very thin specimen, say 100 nm thick, energies of the order of 50–100 keV are needed. It should, however, be clearly appreciated that the diffraction geometry itself does not necessitate such high energies.

In this connection it should be noted that if low electron energies are used so that the penetration is very small,† or if the specimen is *very* thin, then in either case the crystal will act as a two-dimensional grating. The Bragg formula is then not applicable since a diffracted beam will be obtained for *any* angle of incidence (as in an ordinary diffraction grating).

2.17. Neutron diffraction

For neutrons the situation is quite different. Their interaction with most atoms is very weak and they penetrate materials quite readily. Although the diffracted beams may not be very intense, they are, unlike electrons, able to escape from the sample very easily and can then be detected. If we equate the minimum energy for diffraction, $\sim 3.3 \times 10^{-22}$ J, to the thermal energy, $\frac{3}{2}kT$, then this corresponds to the mean thermal energy at a temperature of about 25 K. Neutrons in thermal equilibrium with a system at room temperature, say 300 K, will have a wave vector about three times the reciprocal lattice basic vector (since their energy varies as k^2) and so they will give several orders of diffraction and these will be sufficient for structural analysis. The usual source‡ of thermal neutrons are the slow neutrons in the moderator of a nuclear reactor. These are allowed to escape from the moderator through a hole in the wall of the reactor and they are then collimated and are passed through a velocity selector so that they are mono-energetic, before being used as a beam in neutron diffraction investigations.

2.18. The choice of a diffraction technique

Since the general form of the diffraction pattern will be the same whether X-rays, electrons, or neutrons are used, we must briefly discuss which diffraction technique is the most satisfactory for the various types of investigation.

† In this case the diffracted beam is reflected from the surface. This is a very useful technique for surface-structure studies.

‡ Large numbers of neutrons are also produced when high energy protons (from an accelerator) collide with uranium nuclei. This is called *spallation*. This process has been developed into another technique for providing a source of neutrons. These can be moderated to low energies by water or liquid hydrogen.

In general X-ray diffraction is the cheapest and the most convenient method and is by far the most widely used. X-rays can be detected photographically or with a counter and the latter enables the data to be recorded automatically in digital form for computer input. X-rays are not absorbed very much by air and so the specimen need not be in an evacuated chamber. They do have the disadvantage, already mentioned, that they do not interact very strongly with the lighter elements.

Electrons are scattered strongly in air so that diffraction experiments must be carried out in a high vacuum and whilst nowadays this is a standard routine, it is still an extra complication and expense. Because they are charged particles, electrons interact strongly with *all* atoms and, as we have already discussed, this necessitates a beam of very high energy (50 keV to 1 MeV) as well as specimens which are tiresomely thin (100–1000 nm). The charge on the electron is, however, by no means entirely a nuisance because unlike X-rays and neutrons, electrons can be focussed into narrow beams by electrostatic or magnetic lenses. This enables the diffracted beams to be used to form a direct magnified image of the structure, as in the electron microscope.

Neutron diffraction is a very special tool. A high-flux nuclear reactor or a spallation source is necessary to provide a suitable intense beam of thermal neutrons and the number of such installations available in the entire world is extremely limited. Clearly the investigations which involve neutron diffraction need to be very carefully selected. Among the most straightforward are those in which light atom positions are required. A second group of investigations relies on the fact that, although uncharged, the neutron has an intrinsic magnetic moment and this means that it will interact strongly with atoms or ions in a crystal which also have magnetic moments. Any special regularities in the alignment of the magnetic ions in a crystal can give rise to special 'magnetic' diffraction lines. For this reason neutron diffraction has been a very important tool in the investigation of the magnetic ordering that occurs in some materials, particularly those that become antiferromagnetic (see Chapter 12).

2.19. Inelastic scattering

In all the preceding examples in this chapter we have implicitly assumed that the energy of the emergent radiation is the same as that of the incident radiation; i.e. the waves are scattered *elastically* in the material so that there is no exchange of energy with the atoms in the crystal. However, in some circumstances an energy interchange can occur and this gives rise to *inelastic* scattering. This is a particularly important effect when neutrons pass through a crystal and many neutron diffraction experiments exploit this phenomenon. These inelastic scattering experiments are especially useful in studying the details of the thermal vibration of atoms in a crystal (i.e. the phonon spectrum). A treatment of inelastic scattering, however, is outside the scope of this book.

PROBLEMS

2.1. The six atoms A shown are part of a simple cubic two-dimensional lattice which has a unit cell of side 0·3 nm. Calculate the angles for Bragg scattering at the horizontal planes for X-rays of wavelength of 0·2 nm.

    A   A   A

    A   A   A

By summing the individual contributions to the diffracted beam from each of the six atoms, compare the intensities for the two lowest orders of diffraction (n.b. intensity is proportional to (amplitude)^2).

2.2. In the figure for problem 2.1 another six atoms B are set mid-way between the A atoms, as shown, (A–A remains 0·3 nm). Again, by direct summation for the twelve atoms compare the intensities of the diffracted rays for the same two angles as before, (a) if the scattering power of the B atoms is the same as that of the A atoms, and (b) if their scattering power is one-half of that of the A atoms.

    A   A   A
     B   B   B
    A   A   A
     B   B   B

2.3. The unit cell dimension of f.c.c. copper is 0·36 nm. Calculate the longest wavelength of X-rays which will produce diffraction from the close-packed planes. From what planes could X-rays of wavelength 0·50 nm be diffracted?

2.4. A simple cubic lattice has an atomic spacing of 0·35 nm. Draw a section of its reciprocal lattice parallel to a cube face. The crystal is illuminated by X-rays of wavelength 0·25 nm. Show on your diagram the directions of the possible incident and diffracted beams. What is the total number of different types of reflection?

2.5. The unit cell dimension of b.c.c. iron is 0·29 nm. Two orders of neutron diffraction are observed from the (110) planes. Calculate the minimum neutron energy which is required. At what temperature would such neutrons be dominant, if their distribution is Maxwellian?

2.6. A cubic crystal with lattice spacing 0·4 nm is mounted with its [001] axis perpendicular to an incident X-ray beam of wavelength 0·1 nm. Initially the crystal is set so as to produce a diffracted beam associated with the $(h_1k_1l_1)$ planes. Calculate the angle through which the crystal must be turned in order to produce a beam from the $(h_2k_2l_2)$ planes where

$(h_1k_1l_1) = (020),$ $(h_2k_2l_2) = (030)$

$(h_1k_1l_1) = (020),$ $(h_2k_2l_2) = (130).$

Which, if any, of the $(h_2k_2l_2)$ diffracted beams would be forbidden if the crystal were (a) simple cubic, (b) f.c.c., (c) b.c.c?

2.7. A back-reflection Laue pattern is made of an aluminium crystal (lattice parameter 0·405 nm) using 50 keV radiation. The (111) planes make an angle of 88° with the incident beam. What orders of reflection are present in the beam diffracted by these planes? Ignore diffraction due to wavelengths greater than 0·2 nm which it may be assumed will be too weak to be observed.

2.8. The Bragg angle for a certain reflection from a powder specimen of copper is 47·75° at 20 °C and 46·60° at 1000 °C. Calculate the coefficient of linear expansion of copper.

3. Defects and disorder in crystals

3.1. The influence of defects

IN the preceding chapters we have described the manner in which the crystal lattice is formed by a repeated arrangement of atoms. The details of this arrangement can be determined by analysing the diffraction effects which are produced when a beam of some appropriate radiation (X-rays, electrons, or neutrons) is incident on the crystal.

In the discussion we stressed that our approach is only valid if the atoms are *exactly* at their correct positions in the crystal. This is extremely unlikely when we recall that there are about 10^{22} atoms in 1 cm^3 of material. There must always be some atoms which are not exactly in their right place and so the lattice will contain imperfections or *defects*.

It might be thought that the presence of these defects is merely a mild annoyance which will just slightly affect the properties of a material. It has become clear, however, that many important properties of solids, e.g., the electrical resistance and the mechanical strength, are governed by the presence of certain types of defect in the lattice. It is therefore extremely important that, whilst we must understand the details of the perfect crystal structure, we should also study the ways in which it can be upset and the physical consequences of these defects.

There are two main types of geometrical defect in a crystal. There are those which are very localized and are of atomic dimensions—these are called *point defects*. The most obvious example is an impurity atom. This can be either a *substitutional* or an *interstitial* impurity depending on whether it replaces an atom on the host lattice or whether it is between the host atoms on a non-lattice site. In both cases there will be some distortion around it. There are also those defects which are of a more extended nature—the most important and interesting being the *dislocation*. In addition to these static defects, the perfection of the lattice arrangement will be continually upset by the thermal vibrations of the atoms. This will of course be especially important at high temperatures and we shall need to study what influence the atomic motion has on the properties of crystals.

The presence of defects increases the disorder of the crystal, i.e. it increases its *entropy*. There is therefore a tendency for more point defects to be present at higher temperatures.

3.2. Point defects; vacancies

The simplest defect to consider is the configuration which arises when an atom is missing from its site in the lattice. This vacant lattice is called a *vacancy* (Fig. 3.1) and is sometimes referred to as a Schottky defect. In the region around a vacancy there will be a tendency for the atomic arrangement to readjust slightly and so the crystal lattice becomes distorted.

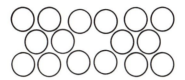

FIG. 3.1. A lattice vacancy (or Schottky defect).

The presence of vacancies provides a means for atoms to diffuse fairly easily from one part of the crystal to another since an atom can move to a vacancy thereby leaving its own site vacant without producing too much disruption of the existing crystal lattice, i.e. very little energy is required. Thus diffusion can be thought of as a migration of vacancies in the opposite direction. In an ionic crystal (section 1.15) the presence of a vacancy at a positive-ion site upsets the electrical neutrality of the region and so the vacancy has an effective negative charge associated with it. This would increase the electrostatic energy of the system; hence in order to maintain electrical neutrality there is a tendency for positive and negative ion vacancies to be produced in pairs.

Not only does the presence of vacancies enhance ordinary diffusion in an ionic crystal but in the presence of an applied electric field the increased diffusion due to vacancies will also assist the electrical conduction.

If E_v is the energy required to create a vacancy (generally about 1 eV) then it can be shown† that for a crystal containing N atoms, there is an equilibrium number n of vacancies at a temperature T which is given by

$$n = N \exp\left(-E_v/kT\right), \qquad (3.1)$$

where k is Boltzmann's constant. A similar expression holds for pairs of vacancies in ionic crystals where n and N are the number of vacancy pairs and ion pairs, respectively, and E_v is the mean of the energy of formation of a positive (E_+) and a negative (E_-) vacancy; i.e. $E_v = \frac{1}{2}(E_+ + E_-)$.

It is clear from (3.1) that the number of vacancies increases with temperature. Changes in vacancy concentration can be detected by measuring the rate of self-diffusion of atoms in a crystal or by measuring the increase in the electrical conductivity of an ionic crystal. Large numbers of vacancies can be 'quenched' in a crystal by heating it (say to within 100 K of its melting point) and then

† By minimizing the free energy of the system, see problem 3.3.

cooling it very rapidly (e.g. by plunging it in water or cold brine) to a temperature at which the vacancies can only migrate and be annihilated very slowly. It is thus possible at room temperature to compare the properties of crystals which contain various concentrations of vacancies (Fig. 3.2).

Materials which contain large numbers of defects, so that the ions can migrate easily (this can occur especially at high temperatures) can have a high ionic electrical conductivity.

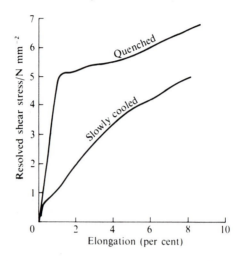

FIG. 3.2. The effect of vacancies on the resolved shear stress of an aluminium single crystal. The extra vacancies in the quenched sample substantially increase the shear stress (R. Maddin and A. H. Cottrell (1955). *Phil. Mag.* (vii) **46**, 735).

3.3. Charge compensation

We have already mentioned that in ionic crystals the ions and also the vacancies always arrange themselves so that no net electrical charge is built up within any small volume of the material. If ions or charges are added to, or removed from, the lattice there will in general be a rearrangement of the ions, or maybe of their outer valence electrons, in order that electrical neutrality is maintained. This rearrangement of charges gives rise to some interesting types of point defect.

Consider what happens if a positive divalent ion is introduced into a monovalent lattice. In order to preserve charge neutrality it will be accompanied by a vacancy on a neighbouring positive-ion site (Fig. 3.3). As an example of this a crystal of K^+Cl^- with $Ca^{2+}Cl_2^-$ admixture will contain more vacancies than pure KCl. Their presence can be deduced from the fact that the mixed crystal has a higher electrical conductivity than pure KCl.

Fig. 3.3. Charge compensation. The presence of the positive ion 2 + in a monovalent lattice encourages the formation of a vacancy (V) so that charge neutrality may be preserved.

3.4. Colour centres

Another set of defects which results from charge compensation are the *colour centres*. If some crystals are irradiated with X-rays, gamma rays, neutrons, or electrons a colour change is often produced, e.g. diamond is coloured blue by electron bombardment and quartz is coloured brown after irradiation by neutrons in a reactor. These colours arise because the irradiation damages the crystal lattice and various types of point defect are produced. In order to maintain charge neutrality in the region of the defect, extra electrons, or maybe positive charges (holes, see Chapter 9) reside at the defects. In the same way that the electrons around an atom have a series of discrete permitted energies so the charges at a point defect also have such a set of levels. These permitted energy levels can be separated from one another by energies corresponding to that of photons in the visible region of the spectrum. Thus light of certain wavelengths can be absorbed at the defect sites, and the material appears to be coloured. In many cases heating the crystal enables the defects to diffuse away so that the irradiation damage is repaired. The crystal then loses its coloration.

A particularly well-studied colour centre is the F-centre (from the German *Farbe*—colour). This is produced by heating an alkali halide crystal in the alkali metal vapour and then quenching it rapidly, e.g. if NaCl is heated in Na vapour it acquires a yellow-brown coloration. When the crystal is hot the Na diffuses into the crystal in excess of the stoichiometric concentration of Na ions. There are no extra Cl^- ions available and so the extra Na atoms will be accompanied by a similar number of vacancies on negative ion sites. In order to maintain charge neutrality the single valence electron from the Na is therefore attracted to the negative-ion vacancy and is trapped there (just as in the perfect crystal it would have been attracted to a Cl atom to form a Cl^- ion). The F-centre (Fig. 3.4) therefore consists of a negative-ion vacancy with a single electron on that site. This has a special set of energy levels which can be investigated spectroscopically and by magnetic-resonance techniques (section 11.11). The F-centre is a trapped-electron centre. It is also possible to produce trapped-hole centres and also small clusters of centres (two or three together) which have different optical properties.

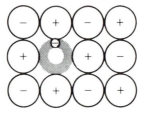

FIG. 3.4. The F-centre. An additional electro-positive atom diffuses into the crystal and is ionized. To preserve charge neutrality the electron which is released on ionization is trapped at a vacancy where a negative ion would normally reside.

3.5. Interstitial atoms

If an extra atom, for which there is no proper lattice site available, is forced into the lattice (Fig. 3.5) then, as in the case of the vacancy, there will be a distortion of the lattice and this defect is usually referred to as an *interstitial*. (If the interstitial originally came from a lattice site, leaving a vacancy, the interstitial–vacancy pair is called a Frenkel defect.) The energy required to produce an interstitial is generally much higher† (five to ten times) than that of a vacancy and so the equilibrium number at any given temperature is usually much smaller. They can however be produced in large numbers by bombardment of a crystal with fast ions or neutrons; they can also be formed during the mechanical deformation of a crystal.

FIG. 3.5. An interstitial atom in the crystal distorts the lattice around it.

3.6. Stacking faults

In section 1.5 we discussed how a face-centred cubic structure may be built up by laying down close-packed planes in a sequence ABCABCA.... A *stacking fault* occurs if this sequence goes wrong, e.g. as in ABCBCABC...,

---

† Except for the silver halides, where the energy is low.

here a layer A is missing (Fig. 3.6(a)), or ABCABACABC... in which an extra A layer has been inserted (Fig. 3.6(b)). Whilst stacking faults can in principle extend through the crystal they usually only occupy part of the plane (Fig. 3.6(c)).

Since in the region of a stacking fault the atoms do not have exactly their correct relationships to one another, there will be some extra elastic strain. This means that there will be an extra surface energy associated with a stacking fault. Note that it is extremely unlikely that a stacking fault of the type AA will occur because the extra elastic energy required for such a strong mismatch would be very large.

A close packed hexagonal lattice of the type ABABA... can also contain stacking faults in which a layer of type C can either be introduced or can be substituted for a layer of type B.

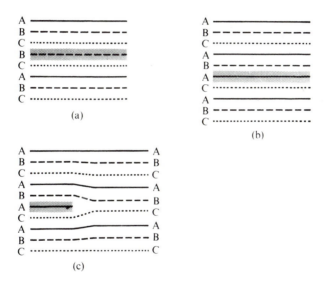

FIG. 3.6. Stacking faults.(a) The regular layering ABCABC... is upset at the shaded region where an A layer is *missing*. (b) A stacking fault formed by the *insertion* of an extra A layer at the shaded region. (c) A stacking fault which terminates within the crystal. The configuration at the termination is called a *partial* dislocation.

3.7. Grain boundaries

Most solids are not single crystals. They are agglomerates of small crystallites randomly oriented with respect to one another. The boundaries between them are known as grain boundaries (Fig. 3.7). In general the structure of a

FIG. 3.7. Crystallites and grain boundaries in α-brass. The specimen was first cold-rolled and then annealed. Twin boundaries are the parallel lines which may be seen in many of the grains. (Magnification × 87.)

grain boundary is exceedingly complex but a special case should be distinguished when the orientation of neighbouring grains is very similar. These are called *low-angle* boundaries. Their geometry is quite simple and can be described in terms of dislocations (section 4.24).

The term *mosaic structure* is sometimes used to describe single crystals in which there are slight changes of orientation between one part of the crystal and another.

3.8. Twin boundaries

Crystals are often produced with a fault in which one region of the crystal is the mirror image of the other. The atoms in one part are in positions produced by reflecting the atoms in the second part at some symmetry plane of the crystal (Fig. 3.8). Twinning often occurs in metals which have a low stacking-fault energy because this implies that the extra energy required for any small atomic mismatch is small. It may also occur during deformation. Twinning planes are often easily observed by optical microscopy (see Fig. 3.7) and the presence of twins can be detected by X-ray diffraction because extra sets of spots are produced from the twinned regions.

FIG. 3.8. Twinning. The crystal structure is reflected in the plane XY which forms the twin boundary. The vertical sides of the crystal, as drawn, are no longer smooth and the boundary can be observed under the microscope as in Fig. 3.7.

3.9. Dislocations

If we consider the types of disorder which can extend beyond the volume of one or two atoms, there are clearly an almost infinite number of configurations. Of these the dislocation is one of the simplest and it is also the most important defect. It is a line defect which can extend right through a crystal or it can form closed loops. Whilst dislocations were originally proposed in order to account for the mechanical strength of crystals, it became apparent that they had a far wider range of influence. Dislocations are necessary, for example, in order to understand the geometry of grain boundaries within crystals and also to account for some of the most important features of crystal growth.

A dislocation is in essence a line defect which can have different atomic arrangements depending on its orientation. Whilst its geometry can be stated in the abstract it is more instructive to discuss why the concept of dislocations was originally introduced, before we describe the actual atomic arrangements which are involved.

The presence of dislocations was suggested independently in 1934 by Taylor, Orowan, and Polanyi in order to account for the observed strength of crystals—particularly metals. Microscopic investigations showed that when a metal crystal is plastically deformed the deformation occurs by the slip of one plane of atoms over another (Fig. 3.9). Deformation does *not* occur by the individual atoms being pulled further away from one another. Depending on the material, the slip occurs on a particular crystallographic plane and in a well-defined direction, which is usually on the plane of closest packing†

---

† e.g. In f.c.c. crystals these are the four equivalent {111} planes. In b.c.c. crystals there are six equivalent {110} slip planes.

along the direction of the line of closest packing in that plane. If a set of crystal specimens are grown with the slip plane at various angles to the axis of tension and the tensile force which produces the first sign of yielding in each of them is measured, then it can be established that the operative stress which produces the deformation is a *shear stress*—i.e. the stress is parallel to the slip plane in the slip direction. This may be shown by calculating the components of the applied force both along the slip direction and also perpendicular to the slip plane (Fig. 3.9(c)). It is the shear stress which is found to be constant for a particular material and it is independent of the orientation of the tensile axis.

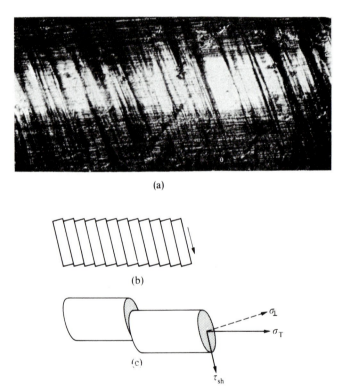

(a)

(b)

(c)

FIG. 3.9. The deformation of crystals. (a) A crystal of cadmium which has been deformed shows slip lines (magnification × 45). These suggest that deformation occurs by the slip of certain planes of atoms over one another as in (b). (c) Enlarged view of a segment of (b). An applied tensile stress σ_T can be resolved into a component, τ_{sh}, parallel to the slip plane in the direction of slip and another $\sigma_\perp$ which is perpendicular to the slip plane. At the yield point τ_{sh} is found to be constant for a particular material. It is independent of the crystal orientation.

3.10. The yield stress by the simultaneous slip of atoms

A crystal may therefore be thought of as deforming like a pack of cards, and on a microscopic scale we might therefore imagine that in order to initiate slip on a particular plane of atoms we must apply a shear stress which is just sufficient to move all the atoms from their original sites on that plane to the next set of equivalent sites one atomic spacing away (Fig. 3.10). Shear strains are measured by the shear angle of displacement α, so that if a is the interplanar spacing and x the linear displacement then $\alpha = x/a$. For a small elastic displacement the shear stress τ is given by

$$\tau = G\alpha = Gx/a, \tag{3.2}$$

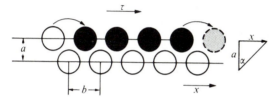

Fɪɢ. 3.10. Calculation of the shear stress required to produce the simultaneous slip of one plane of atoms over another (see text).

where G is the shear (or rigidity) modulus. For larger displacements it is clear that τ must be a function which has the periodicity of the interatomic spacing b of the atoms within the slip plane. For simplicity we shall write it in the form

$$\tau = \text{constant} \times \sin (2\pi x/b). \tag{3.3}$$

But for small values of x (3.2) must hold, and hence we can derive the constant in (3.3) by equating (3.2) and (3.3) for small values of the sine, so that

$$\text{constant} \times 2\pi x/b = Gx/a,$$

hence

$$\text{constant} = Gb/(2\pi a) \approx G/2\pi,$$

since b is approximately equal to a; and so

$$\tau \approx G/(2\pi) \sin (2\pi x/b).$$

Thus τ has a maximum value of $G/2\pi$, and it should correspond to the yield stress of the material, but it does not! The yield stresses of all metals are always very much smaller than $G/2\pi$ by two to four orders of magnitude. More sophisticated calculations than the one which we have presented still give a value of $\sim G/15$ for the yield stress. One is therefore led to the conclusion that the theoretical model is wrong.

The picture we have had of one plane of atoms moving over another like the cards in a pack is not quite correct even though the microscopic observations suggest that this is what does occur. The mistake we have made is to assume that the atoms in one plane move *simultaneously* as a plane over those in the neighbouring layer. It is clear that this could never be quite true. The loading cannot be applied so precisely that the force acting on one atom will be exactly the same as that on the others and the thermal vibration of the atoms will also affect the instantaneous force on a particular atom. So we are led to consider the possibility that some atoms might move before others, i.e. slip might occur by *consecutive* motion rather than by the simultaneous motion of atoms. Dislocations are a necessary consequence of this idea of consecutive motion.

3.11. Consecutive slip of atoms

Consider the situation in Fig. 3.11(a) at an instant during a deformation process in which we now assume that some atoms move before others. Let us assume (Fig. 3.11(b)) that in the shaded region the atoms above the page have slipped by one complete atomic spacing to the right relative to the atoms in the plane of the paper. These slipped atoms therefore will still be in register with the atoms in the plane of the paper. The atoms in the unshaded region have yet to slip and therefore they are also in their correct positions relative to the atoms in the slip plane. But what happens to the atoms at the boundary between the shaded and the unshaded regions? In both regions the atoms are in register throughout the crystal and yet in the shaded area the atoms above the page have all moved one atomic spacing to the right, whereas in the unshaded region they have not yet moved. At the boundary between

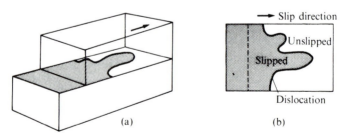

(a)   (b)

FIG. 3.11. The consecutive motion of atoms in the slip process. (a) The top part of the crystal has been sheared in the direction of the arrow. In the shaded region slip by one atomic spacing has occurred between the upper and lower parts of the crystal. (b) A plan of (a) in the slip plane. The boundary between the slipped and unslipped regions is the dislocation line.

the regions the atoms are going to be squashed up in some way and so they will not be in their correct relative positions on either side of the slip plane. It is this line of misplaced atoms between the slipped and the unslipped areas which is called a *dislocation*.

3.12. Edge dislocations

In order to visualize the disposition of the atoms in the region of the dislocation it is useful to consider the particularly simple case of Fig. 3.12. This is when the boundary between the slipped and the unslipped regions is perpendicular to the direction of slip (Fig. 3.12(a, b)). As we continue to apply a stress we may suppose that this boundary moves to the right (Fig. 3.12(b)), until eventually the whole of the upper section of the crystal has slipped over the lower one (Fig. 3.12(c)). The dislocation itself has now vanished—it has passed right through the crystal. We see (lower part of Fig. 3.12(b)) that immediately above the dislocation the interatomic spacing has been decreased whilst below it, it has been increased. In order to accommodate the slip which has occurred to the left of the dislocation an extra half plane of atoms has been effectively inserted into the crystal and this half plane terminates at the dislocation. The atoms at the bottom of this half plane do not have proper partners below the slip plane. The interatomic forces along the dislocation are therefore quite different from the forces in the main body of the crystal—or indeed over the remainder of the slip plane. This type of dislocation is called an *edge dislocation*.

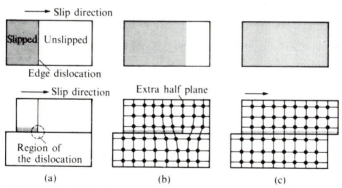

Fig. 3.12. An edge dislocation is formed if the slipped–unslipped boundary is perpendicular to the direction of slip. The upper and lower parts of the diagrams show the plan and side views respectively. (a) During the initial stages of slip. (b) After further deformation the edge dislocation has moved to the right; the atomic configuration around the dislocation should be noted. (c) When slip of the upper over the lower half of the crystal is complete the dislocation has passed right through the crystal and has disappeared. The relative atomic positions are now as they would be in a perfect crystal.

3.13. The yielding of crystals

It will be recalled that the concept of dislocations was introduced in order to explain the relative ease with which crystals can be deformed. Further consideration shows qualitatively how this comes about. The atom A at the centre of the dislocation (Fig. 3.13(a)) has no partner on the lower plane and it will be more or less equally attracted by atoms B and C. Therefore only a small force is required to move it a short distance to the right so that the influence of C becomes dominant. It can then link up with C as a proper partner, thereby forcing atom D to become unpaired (Fig. 3.13(b)). If this happens to all the atoms at A and C in a line normal to the page then the dislocation configuration has moved one atomic spacing to the right—from A to D—and the force required to do this is very small. This mechanism can continue with the dislocation moving further to the right until we have the situation already shown in Fig. 3.12(c) in which it has passed right through the crystal and during which process the whole of one part of the crystal has slipped over the other part.

This method of producing slip by the consecutive motion of atoms can be illustrated by a very illuminating analogy which is now widely used, but which I believe was first suggested by Mott. Consider the easiest way of moving a heavy carpet over a floor. We can drag the whole carpet over the floor all at once (simultaneous slip) and this will require a large force. On the other hand, we can make a ruck at one end of the carpet and then the ruck can be made to run to the other end with only a small applied force (consecutive slip). For further slip we make another ruck and run it through the carpet, and so on. When the ruck has reached the end of the carpet it vanishes and the carpet is perfectly flat again. A dislocation can be considered as a ruck in the atomic carpet on the slip plane. It moves very easily and when it has passed from one end of the crystal to the other, slip equal to the size of the ruck (i.e. the atomic spacing) has been produced.

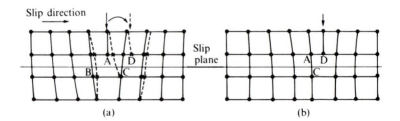

FIG. 3.13. The motion of an edge dislocation is caused by the slight displacement of atom A at the centre of the dislocation (a) so that it pairs up with atom C. This leaves D unpaired (b) and hence the dislocation has moved one atomic spacing to the right.

3.14. Screw dislocations

The edge dislocation which has just been described is the one which is the easiest to envisage. But there is another equally simple situation which should be discussed. This occurs when the boundary between the slipped and the unslipped regions of the crystal lies *parallel* to the direction of slip (Fig. 3.14(a, b)). Unfortunately it is a little more difficult to picture the atomic configuration at this dislocation. It has no extra half plane of atoms but, instead, the atoms along the line of the slipped–unslipped boundary rearrange themselves so that they lie on a helix whose axis is at the centre of the boundary line (Fig. 3.14(c)). It is therefore called a *screw dislocation*. Note how the presence of a screw dislocation turns the separate planes of a crystal into one single plane which is wound up into a spiral ramp (Fig. 3.14(d)).

A dislocation which lies in some arbitrary direction to the slip vector can be resolved into two components which have edge and screw character respectively. This is called a *mixed dislocation*.

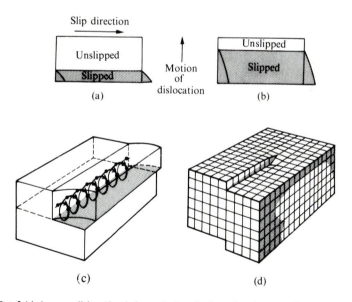

FIG. 3.14. A screw dislocation is formed when the boundary between slipped and un-slipped regions is parallel to the direction of slip. A plan view of the slip plane at two instants during slip is shown in (a) and (b). The atomic configuration around a screw dislocation (c) is in the form of a helix. (d) Shows that the presence of a screw dis-location converts the separate atomic planes into a continuous spiral ramp. If you start on the top plane and continue to make circuits around the dislocation you eventually end on the bottom plane without having jumped from one plane to another.

From our definition of a dislocation as the boundary between slipped and unslipped regions of the crystal, it is clear that a dislocation must either end on the free surfaces of the crystal or it must form a closed loop within the material.

3.15. The mechanical properties of crystals

The main purpose of introducing the concept of dislocations at this stage is to emphasize how the presence of a defect can completely dominate certain physical properties of crystals, but it would be out of place in this chapter to discuss in detail the way in which dislocation theory is able to account for the plastic deformation of crystals. To do this one needs to take account of those mechanisms which can impede the motion of dislocations. In addition we have already noted that during deformation dislocations pass right through a crystal and thereby vanish. For continuing deformation we therefore require a mechanism for the *multiplication* of dislocations. A further development of dislocation theory and a description of some of the remarkable experiments which clearly demonstrate the existence of dislocations is given in Chapter 4.

3.16. The thermal vibrations of the lattice

However carefully we prepare a specimen, using extremely high purity materials and very specially controlled methods of crystal growth, one form of defect will always be present—that due to the thermal vibration of the atoms. At any instant in time the atoms are never exactly at their correct lattice sites. At room temperature they are vibrating with approximately simple harmonic motion at around 10^{13} Hz about an origin which is at the geometrical lattice position. Even at very low temperatures the zero-point motion of the atoms is still present. As the temperature is raised the amplitude of the atomic vibrations increases and this means that virtually every property of a material whose magnitude is determined by the actual position of the atoms, or by the thermal energy, changes with temperature.

An example which we have already discussed is the variation in the intensity of X-ray patterns (section 2.12) and the reader will also be aware of the increase in the electrical resistance of metals as the temperature is raised. Occasionally the effect of the atomic vibrations is countered by some other mechanism which changes with temperature in such a way that the final behaviour is temperature independent. One example of this is the constant value of the thermal conductivity of metals at and above room temperature.

In this chapter we introduce the idea of atomic vibrations as a type of lattice defect which can give rise to temperature dependent properties. It should be appreciated, however, that there are other characteristics of solids which are dependent on the way in which the thermal energy of a system varies with temperature but in which the lattice disorder itself is not an important factor. Examples of this are the variation in the concentration of current carriers in semiconductors and the change in the specific heat of all materials with temperature.

When is it necessary to take account of the actual disorder produced by the thermal vibrations? In general terms this will be when the time scale of the interacting system is short compared with the period of the lattice vibrations. In that case it is only the instantaneous position of the atoms which is important and hence their random positions must be considered. Thus an electron which moves in a metal at about 10^6 m s^{-1} passes the atoms which are spaced 10^{-10} m apart in 10^{-16} s. This is much smaller than 10^{-13} s, the period of an atomic vibration and therefore, so far as an electron is concerned, the atomic displacements cannot be averaged out to zero and the atomic vibrations will contribute to electron scattering.

A more detailed mathematical treatment of lattice vibrations and thermal energy is given in Chapter 5.

PROBLEMS

3.1. Point defects in metals can cause additional electrical resistivity at low temperatures due to extra electron scattering which is proportional to the number of defects. The table gives the relative change in the resistivity at 78 K of a gold wire when it is quenched from various temperatures.

Temp K	920	970	1020	1060	1220
Resistivity change	0.41%	0.7%	1.4%	2.3%	9.0%

Calculate the energy of formation of a vacancy in gold.

3.2. The energy of formation of a vacancy in copper is 1 eV. Estimate the relative change in the density of copper due to vacancy formation at a temperature just below its melting point, 1356 K.

3.3. The number of different ways of arranging n vacancies on N atomic sites is $N!/\{(N-n)!n!\}$. If the energy of formation of a vacancy is E_v. calculate the extra free energy due to the presence of vacancies. By minimizing this free energy show that the equilibrium number of vacancies at a temperature T is given by $n \approx N \exp(-E_v/kT)$. (Use Stirling's approximation

$$\ln N! \approx N \ln N.)$$

3.4. When a dislocation passes right through a crystal slip equal to a lattice spacing occurs. Estimate the number of dislocations on the (111) plane which would be needed to produce 5 per cent plastic deformation in an aluminium specimen 1 cm long. (For Al, cube edge of lattice = 0.4 nm.)

3.5. Draw the cubic unit cell of a f.c.c. metal and indicate the main slip planes and the slip directions. How many equivalent slip planes and directions are there?

3.6. A long crystal of cross-sectional area A is subjected to a tensile force F along its axis. Show that if θ is the angle between the tensile axis and the normal to the slip plane and ϕ is the angle between the slip direction and the tensile axis, then the shearing stress in the slip plane is equal to $(F/A) \cos \theta \cos \phi$.

3.7. A metal single crystal in the form of a cylinder 1 cm in diameter has its axis at 30° to the normal to the slip plane. The direction of slip is at 45° to the cylinder axis. If the stress to produce shear in the slip plane is $10^6 \, \text{N m}^{-2}$ calculate the force which must be applied to the cylinder axis in order to produce yielding.

3.8. If the angles of the unit cells shown in Fig. 3.8 are about 89° and 91° sketch an experimental arrangement by which the angle of the twin boundary could be determined accurately.

3.9. If it is assumed that the strain due to a dislocation is homogeneously contained within a region of radius five atomic spacings around the line of an edge dislocation, make a rough estimate of the elastic energy of 1 m of edge dislocation in copper. (For Cu, bulk modulus $= 1.4 \times 10^{11} \, \text{N m}^{-2}$; interatomic spacing $= 0.25 \, \text{nm}$.)

3.10. The edge dislocation in Fig. 3.13 is shown moving in a horizontal plane. Explain why it would be difficult to move it in a direction perpendicular to that plane. In what planes would you think that a screw dislocation could move easily?

3.11 The formation energy of a vacancy in Al is 0.75 eV. If Al is rapidly quenched from a high temperature to 300 K, estimate the average lifetime of a vacancy, assuming that it is annihilated when an atom diffuses to the vacancy site. The energy for self-diffusion of an atom (i.e. the energy to form a vacancy and move to a vacancy site) is 1.5 eV. Atoms and vacancies vibrate at $\sim 10^{13}$ Hz. Assume that each vacancy travels, on average, 300 nm from its original site. The atomic separation in Al is 0.29 nm. [N.B. This is random hopping motion and so the distance travelled is proportional to (number of hops)$^{1/2}$.]

4. Dislocations in crystals

WE have already introduced the idea of a dislocation as being a line defect which separates the slipped and the unslipped regions of a crystal which is being deformed (section 3.9 *et seq.*). In this chapter we develop some of these ideas and we discuss the scope of dislocation theory.

4.1. Dislocation density

The number of dislocations in a specimen is clearly an important quantity. We define the dislocation density N_d as the total length of dislocation line in unit volume. It is also equal to the number of dislocations which cut through a unit area which is randomly oriented in the crystal.

In annealed crystals N_d is of the order of 10^{10} m^{-2}; in work-hardened materials it can be as high as 10^{16} m^{-2}.

4.2. The Burgers' vector

A dislocation can either extend right through a crystal or it can form a closed loop. Therefore it cannot be described just in terms of its orientation since this can vary with position. The entire dislocation can be characterized however by a vector which represents the amount and direction of slip which is produced when that dislocation has passed right through the crystal (in a certain direction to be established by convention). This vector is called the Burgers' vector **b**. Its magnitude (sometimes called the *strength* of the dislocation) is a repeat vector of the lattice and it is usually the smallest one, so that it is of the order of the interatomic spacing. From the definition of a dislocation it should be clear that its Burgers' vector is constant throughout its length even though the orientation of the dislocation itself may change. The Burgers' vector lies in the slip plane and it will be perpendicular to an edge dislocation and parallel to a screw dislocation.

4.3. The conservation of the Burgers' vector

Dislocations may join up with one another to form a series of networks within the crystal. If a dislocation splits into a number of other dislocations then the sum of the Burgers' vectors of the components must equal the Burgers' vector of the original dislocation.† This is of course an analogue of Kirchhoff's law for electric currents at a node.

---

† A consistent sign convention must be used. See the following section.

4.4. Dislocations of opposite sign

In Fig. 3.12, we drew an edge dislocation with its extra half plane at the top of the crystal. We could clearly have another very similar dislocation in which this half plane was in the lower half of the crystal. The slip produced by this second dislocation if it passed through the crystal *in the same direction as the first* would be in the opposite direction. Their Burgers' vectors would have opposite signs.

If such a pair of dislocations meets on the same slip plane they will annihilate one another because the two extra half planes will join up to form a perfect lattice (Fig. 4.1)—this is also borne out by the fact that the sum of their Burgers' vectors is zero. The positions of the extra half plane relative to the slip plane are shown by the conventional signs ⊥ or ⊤, for edge dislocations.

Similar reasoning shows that we can have either right- or left-handed screw dislocations (section 3.14, p. 49) depending on the sense of the helix. A pair of these will also combine to form a perfect crystal.

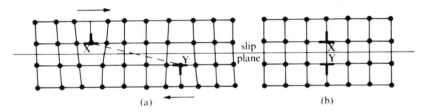

FIG. 4.1. Two edge dislocations X, Y of opposite sign on the same slip plane (a) are attracted to and annihilate one another (b) thereby leaving a perfect crystal with a consequent reduction in the elastic energy of the system.

4.5. Impurities and dislocations

At an edge dislocation the atoms are closer together just above the slip plane where the extra half plane stops (Fig. 4.2), and therefore any impurity atom A which is smaller than the host atoms can be accommodated in this part of the crystal with a reduction in the elastic energy. On the other hand, a large atom B will be attracted to the other (lower) side of the slip plane where the lattice is extended. Thus edge dislocations act as a place where impurity atoms will tend to be concentrated. Reference is often made in the literature to the *impurity atmosphere* around a dislocation. For a similar reason dislocations on parallel planes will tend to line up under one another because the elastic energy around them is thereby reduced.

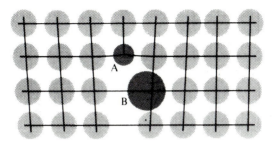

FIG. 4.2. Both small (A) and large (B) impurity atoms can be accommodated more easily around an edge dislocation than in the perfect crystal. Impurities therefore tend to cluster along dislocation lines.

4.6. Conservative and non-conservative motion

An edge dislocation can move easily in either horizontal direction along the slip plane. This is called *conservative* motion because when the dislocation passes through a region the number of lattice sites remains constant. If, however, the dislocation was made to move vertically upwards some of the extra half plane would have to be dismantled, and so a trail of interstitial atoms would be left behind (Fig. 4.3). Similarly if the motion was downwards then the extra atoms which would be needed to extend the extra half plane would result in vacancies being produced. Both these types of motion are *non-conservative* and are referred to as dislocation climb. Because extra energy is required to produce the point defects, dislocation climb only occurs at high temperatures. Pure screw dislocations do not have an extra half plane and they can move conservatively on any slip plane.

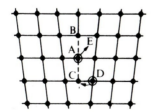

FIG. 4.3. Climb of an edge dislocation. An edge dislocation can only move easily within its own slip plane. If it 'climbs' out of the slip plane from A to B then A must become an interstitial E. If it moves from A to C then there must be an atom at C which must come from a site such as D, thereby leaving a vacancy at D.

4.7. Stresses and strains around a dislocation

Whenever a lattice is distorted from its regular structure extra elastic energy is associated with the distortion. This is so because the original structure must have been the one with the lowest (free) energy; if it were not, it would not have been the original structure! We need to know the extra energy which is associated with the presence of a dislocation and to calculate this we require a detailed knowledge of the way the lattice is strained in that region. This is fairly easy to do for a screw dislocation.

In Fig. 4.4 we have drawn the cylindrical region of material around a screw dislocation. If we exclude the central core of atoms in the immediate vicinity of the helix, there is at any radius r from the centre a displacement equal to the Burgers' vector b, around a complete circumference. The shear strain $\varepsilon_{\theta z}$† at any point is therefore

$$\varepsilon_{\theta z} = b/(2\pi r). \tag{4.1}$$

It will be noted that since this strain only falls off as r^{-1} it is a *long-range* strain. Hence while we tend to describe a dislocation as a *line*, its influence is felt some distance away from the central region.

Associated with the strain (4.1) there will be a shear stress $\tau_{\theta z}$, which, assuming that we can use elasticity theory, will be given by

$$\tau_{\theta z} = Gb/(2\pi r), \tag{4.2}$$

where G is the shear modulus.

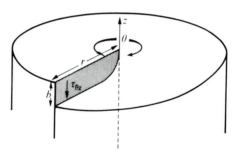

FIG. 4.4. The strain around a screw dislocation (see text). Note that the *displacement* of the atoms, b, is independent of the distance r from the centre of the dislocation (except very close to the dislocation core). The shear stress at any point is given by $\tau_{\theta z}$.

---

† The convention is that the first suffix is the direction of the normal to the shear plane and the second suffix is the direction of shear. τ is used for shear stress and σ for normal stress.

The elastic energy per unit volume is given by $\frac{1}{2}$ stress × strain and so the energy of unit length of the cylinder shown in Fig. 4.4 will be

$$E = \frac{1}{2} \int_{r_0}^{R} \frac{Gb}{2\pi r} \frac{b}{2\pi r} 2\pi r \, dr,$$

where R is the outer radius of the crystal and r_0 is the radius of the core of the dislocation. Thus

$$E = (Gb^2/4\pi) \ln (R/r_0). \qquad (4.3)$$

In order to go beyond (4.3) we need to consider both R and r_0. If the crystal does contain just one dislocation (which is unlikely) then R will be equal to the radius of the crystal. If, as is more usual, many dislocations are present, then their arrangement is such that the stress fields of neighbouring dislocations usually tend to oppose one another. The value for R can then be reasonably taken to be half the distance between neighbouring dislocations. Eqn (4.3) does not include the contribution from within the core of the dislocation ($r < r_0$). This is difficult to calculate because in this region the material cannot be treated as an elastic continuum. The strains are so large that we must have a knowledge of the actual atomic positions and any division between the core and the elastic region is obviously arbitrary. It is a problem which has not yet been properly resolved since it depends on non-linear force laws.

The calculation of the energy of an edge dislocation is more complicated because both shear and normal stresses are present. A calculation which neglects the core of the dislocation can be made yielding an expression which is similar in form to (4.3).

To a first approximation the energy per unit length of a dislocation, independent of its character, is usually assumed to be

$$E \approx Gb^2. \qquad (4.4)$$

If G is taken to be 4×10^{10} N m^{-2} and b is $2 \cdot 5 \times 10^{-10}$ m then the energy of a dislocation is about $2 \cdot 5 \times 10^{-9}$ J m^{-1}, or about 4 eV for each atomic plane threaded by the dislocation.

4.8. The magnitude of the Burgers' vector

In all our discussion we have assumed that the slip associated with a dislocation has been equal to *one* lattice spacing. It might well be asked why in any of the figures (e.g. Fig. 3.12) should not the region which has already slipped do so by two or more spacings with a consequent increase in the magnitude of the Burgers' vector for the associated dislocation. From (4.4), however, we see that since the energy of a dislocation is proportional to b^2, it is energetically more favourable for a dislocation with a large value of b to separate into several dislocations each of which has as small a value of b as possible. In many cases this means that b is equal in magnitude to the lattice spacing a, although in some structures it is possible to have *partial dislocations*† in which b is less than a. The tendency to minimize the value of b implies that there will be a repulsion between like dislocations which are on the same slip plane.

4.9. The line tension of a dislocation

Since there is an additional elastic energy E associated with unit length of dislocation (4.4) there will always be a tendency for a dislocation to make itself as short as possible so as to minimize this energy. The dislocation behaves like a piece of elastic; closed loops will therefore tend to contract and disappear and other dislocations will become as straight as possible.

If the length of a dislocation is increased by δx then work equal to $E\,\delta x$ must be done. This can be thought of as being produced by a tensile force T moving through the distance δx, and hence $T = E$. The dislocation, therefore, can be considered to have a line tension T equal to $E \approx Gb^2$. This is of course completely analogous to the surface tension in a liquid surface which is equal to the surface energy per unit area. A straightforward calculation shows that in order to bend a dislocation into an arc with a radius of curvature R requires a normal force F per unit length given by

$$F = T/R \approx Gb^2/R. \tag{4.5}$$

4.10. The force on a dislocation

Although we deform a crystal by applying an external shear stress to it, we have described the mechanism of deformation in terms of the motion of dislocations. These must therefore experience a force when a stress is applied to the crystal otherwise they would not move. Simple considerations enable us to derive an expression for this force.

When a dislocation of Burgers' vector b sweeps over the entire area A of a slip plane, slip equal to b is produced. If *part* of the dislocation of length l moves a distance δx over the slip plane (Fig. 4.5) thereby sweeping out an area $l\,\delta x$, a proportionate amount of slip $(l\,\delta x/A)b$ will result. But this slip has actually been produced by an external shear stress τ acting on area A. The work done by this stress is force × distance which is equal to $\tau A(l\,\delta x/A)b$. However, we could ignore the external stress and say that the deformation was produced as if the dislocation was being driven by a force F per unit length through the distance δx. The work done must be the same in both cases and so we can equate

$$F l\,\delta x = \tau A(l\,\delta x/A)b,$$

and so

$$F = \tau b. \tag{4.6}$$

† For example, see Fig. 3.6(c).

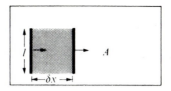

FIG. 4.5. Calculation of the force on a dislocation, produced by an external stress (see text).

This is the force experienced by unit length of dislocation when an external stress τ is applied to the crystal.

4.11. The force of one dislocation on another

If two dislocations are close together the stress field of one—$Gb/(2\pi r)$—(4.2) will act on the other just as if it were an external stress. Combining (4.2) and (4.6) gives an expression for the radial force F_{d-d} between two screw dislocations:

$$F_{d-d} = Gb^2/(2\pi d) \quad \text{per unit length,} \tag{4.7}$$

where d is the distance between the dislocations.

4.12. The multiplication of dislocations

Many crystals can double their length during tensile deformation. If we have a specimen 1 cm long with a 1 mm square cross-section and each dislocation can produce a deformation of $b \sim 2.5 \times 10^{-10}$ m, the number of dislocations which are necessary to produce an extension of 1 cm would be about 4×10^7. The area of the side of the specimen is 10^{-5} m^2 and hence we would require a dislocation density of 4×10^{12} m^{-2} to produce this deformation. Now annealed crystals only contain about 10^{10} dislocations per square metre, and thus they could produce a maximum deformation of less than 1 per cent. We are led to the conclusion therefore that extra dislocations must be produced during deformation.

4.13. The Frank–Read source

This is the simplest mechanism which seems to actually occur for the multiplication of dislocations. We consider a length of dislocation line which is firmly anchored at two points A and B (Fig. 4.6(a)). The anchoring may be achieved by impurities, by nodes in a dislocation network, or by some crystal defect. If a stress is applied to the crystal we have seen (section 4.10) that this will exert a force on the dislocation and so it will tend to bow out (Fig. 4.6(b)). As the dislocation moves the area which it sweeps out above the slip plane shifts by b relative to the area below the slip plane. Eventually the dislocation

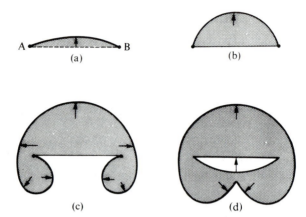

FIG. 4.6. The Frank–Read source. A dislocation which is pinned at A and B is bowed out (a), (b), by an applied force. This produces some slip (shaded regions). As the process continues lobes are formed (c) which ultimately touch (d) thereby recombining to form perfect crystal. There remains a closed dislocation loop and the original pinned dislocation AB. The mechanism can be repeated to form further loops.

becomes a semicircle and its radius of curvature cannot decrease any further. As the stress is increased, lobes begin to form and these will eventually touch (Fig. 4.6(c), (d)). Now here is the crucial point. The two lobes are parts of one dislocation which are moving in opposite directions and yet they must produce the *same* deformation. Therefore they must be of opposite sign and since they are on the same slip plane they will annihilate one another (section 4.4) to form a piece of perfect crystal. They leave behind them a closed dislocation loop *and* the original anchored section of dislocation (Fig. 4.6(d)). This process can be repeated so that eventually a whole series of loops are formed one within the next.

As more inner loops are formed the outer ones expand thereby producing extra slip. The process stops when the outer loop encounters a barrier which prevents further expansion. Since it repels like dislocations on the same plane (section 4.8) the barrier will exert a back stress on all the inner loops. This will oppose the applied stress and so eventually the source will cease to operate.

4.14. The strength of materials; work-hardening

When a stress is applied to any material the general pattern of behaviour is as follows. Initially there is a small amount of *elastic* strain, which is usually only a fraction of 1 per cent. If the stress is removed the specimen returns to its original length. It is of course the behaviour of the material in this region which defines the various elastic moduli of the material. As the stress is increased the specimen *yields* and this is followed by *plastic* deformation (Fig. 4.7). The amount of plastic flow can vary from virtually zero (in brittle materials) to more than 100 per cent. When the stress is removed the length decreases only very slightly and if it is then reapplied there is first a new short elastic region and then plastic yielding starts again at the stress to which the specimen was previously loaded. The material has therefore been hardened by the initial amount of plastic deformation. This is called *work-hardening*.

The strength of a material can therefore be either an intrinsic property for which an understanding of the original yield stress is desirable, or, if it has been work-hardened, we need to know the mechanisms which operate during its original plastic deformation.

A study of the tensile properties of materials can be made from two points of view: (a) either to understand, and possibly to control, the properties of constructional materials; or (b) to investigate the fundamental properties of special laboratory specimens.

Constructional materials are used under conditions of stress in which they must *not* yield (or at most only to a very carefully regulated degree). We are therefore interested in mechanisms which control the yield stress and which can raise it to as high a value as possible. Nevertheless the processes which occur *during* plastic deformation, whilst of little concern to the constructional engineer, are still very important in the *fabrication* of materials, e.g. in pressing out a car body or in a rolling mill.

Fundamental studies, particularly on single crystals, tend to concentrate on processes which occur during the plastic deformation, rather than with the yield stress itself; these processes, however, are of more than academic interest because, as mentioned above, a material can be work-hardened by plastic deformation and its yield stress is thereby raised.

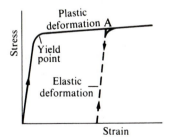

FIG. 4.7. A typical stress–strain diagram for a material which deforms plastically. If the stress is relaxed to zero at A and is then re-applied there is a region of elastic deformation (dotted line) followed by a yield at A. This increase in the yield point from its original value to that at *A* is called *work hardening*.

4.15. The yield point in annealed crystals

The initial yielding of a crystal occurs when a stress is applied which is sufficient to start the dislocations moving freely. In otherwise-perfect f.c.c. and h.c.p. crystals this stress (known as the Peierls–Nabarro stress) is very small† ($\sim 10^5$ N m^{-2}). Unless the atoms are 'hard' and the lattice is very rigid, so that all the distortion due to the dislocation occurs in its immediate vicinity, rather than being spread out over a wider region, the Peierls–Nabarro stress is *not* the controlling factor for the yield point. Indeed, in most materials, the dislocation is locked in position by impurity atoms or other defects‡ and the initial yielding is determined by the stress which is required to move it clear of these obstacles. We now discuss some mechanisms in which the yield point is affected by impurity atoms.

4.16. Hardening by impurity atoms

Impurity atoms can be dispersed in a crystal in a number of ways, each of which will interact with dislocations in a different manner and give rise to a different hardening mechanism. The most important of these are as follows:

(1) The impurities may be dispersed as individual atoms and occupy lattice sites (substitution) or interstitial sites. In either case, as we have already seen (section 4.5, p. 54), they will interact with dislocations because their presence will change the elastic energy of the configuration. In order to move a dislocation through an impurity region extra energy is required. This is the basis of *solid solution* hardening. An example of substitutional hardening is that of tin in copper (bronze); interstitial hardening can be produced by small amounts of interstitial carbon in iron.

(2) The impurity atoms may agglomerate as a result of various heat treatments to form precipitates. These may consist either of the impurity atoms alone, or of a compound of the impurity with the host material. Their size, depending on the alloy, might range from 20 nm to 500 nm.

The presence of the precipitates can impede the motion of dislocations (and hence produce hardening) in two ways. The dislocations might be dragged through the precipitate, or, if the stress to do this is too high, then the dislocation will be anchored at the precipitates which will act as pinning points. As the stress is increased the dislocation will bow out between the precipitates (Fig. 4.8) until loops form round them in a manner similar to that proposed for the Frank–Read mechanism for dislocation multiplication (section 4.13).

---

† It is high for covalent (e.g. diamond. Ge, Si) and complex structures, and also for b.c.c. at low temperatures.

‡ Or by the length of the longest Frank–Read source.

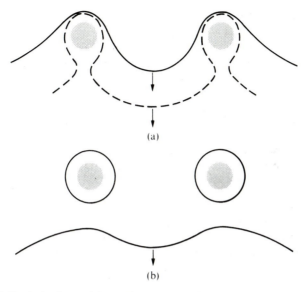

FIG. 4.8. Hardening by precipitates (Orowan hardening). (a) A dislocation which is held up by precipitates is bowed out by an applied force. Lobes are formed on the dislocation and these join up (b) leaving closed loops around the precipitates and a dislocation which has passed clear of them. The process is similar to that of the Frank–Read source (Fig. 4.6).

In general the larger the number of precipitates the more effective will be the hardening by either of the above mechanisms. But a fine dispersal of very small precipitates will not usually be very effective because the dislocations will be able to cut through each one quite easily. Careful heat-treatment followed by quenching is necessary in order that the size and distribution of precipitates is such that materials with optimum properties are produced. An important example of precipitate-hardening is that of aluminium by copper. This is the basis for many lightweight, high strength alloys.

If the precipitates are allowed to grow too large the mechanical properties deteriorate and the alloy is said to be *over-aged*.

The yield point in work-hardened materials

4.17. The main types of stress–strain curves

Typical stress–strain curves for f.c.c., h.c.p., and b.c.c. single crystals (Fig. 4.9) show that after yielding, each type of crystal exhibits a distinctive behaviour.

The f.c.c. curve can be divided into three stages, usually called stages 1, 2, and 3. Stage 1 begins immediately after yielding and is called the easy glide region because the slope of the curve is small, of the order of $10^{-3}G$, i.e. there is very little work-hardening. This is followed by stage 2 which is a linear region in which the work-hardening is substantially increased (slope $\sim 10^{-2}G$). Eventually the curve bends over to stage 3, which is approximately parabolic and in which the work-hardening becomes smaller. In polycrystals stage 1 is not present and the behaviour after yielding is similar to the end of stage 2, continuing into stage 3.

H.c.p. crystals (Fig. 4.9(b)) only exhibit stage 1 behaviour. B.c.c. crystals (Fig. 4.9(c)) often show the phenomenon of the *yield drop*, in which the stress falls immediately after yielding. This is followed by a region of work-hardening. The yield drop is usually more accentuated in polycrystals.

The dislocation processes which occur in these various stages are of course very complicated and indeed, they have not all been completely established. To a certain extent their relative importance will depend on the particular material which is being investigated. We can therefore only give a qualitative, and sometimes tentative, explanation for the phenomena. We shall concentrate on the f.c.c. curve.

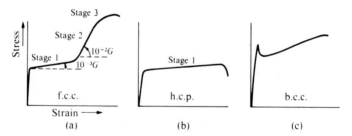

FIG. 4.9. Stress–strain curves for (a) face-centred-cubic, (b) hexagonal-close-packed and (c) body-centred-cubic crystals. Note the three stages of deformation in (a), the single stage in (b) and the yield drop in (c).

4.18. Stage 1—easy glide

In the easy-glide region for f.c.c. and h.c.p. crystals the dislocations can move fairly freely over large distances. Slip occurs on one set of parallel planes. Now we have already seen (section 4.5, p. 54) that there is a tendency for dislocations on parallel planes to line up under one another and in (4.7) we have calculated the force F_{d-d} between two screw dislocations. An extra stress will therefore be necessary to overcome F_{d-d} when a dislocation on one plane moves past another one on a parallel plane. A calculation using (4.7) shows that the effect of F_{d-d} could account for the shear stresses which are involved in stage 1.

As deformation proceeds during stage 1 there is some dislocation multiplication, they start to interact with one another, and so the stress for further deformation continually increases. Towards the end of stage 1 some slip and dislocation multiplication would have occurred on other sets of glide planes which are not parallel to the primary glide plane. Some of these dislocations, as they get close to or cross the primary glide plane will tend to hinder the motion of the original dislocations. This becomes more pronounced as stage 2 is entered. In h.c.p. crystals there is only one main glide plane, the hexagonal basal plane; the effect of slip, and hence dislocations, on other intersecting planes is then not very pronounced. The crystal therefore continues to deform in stage 1 until it fractures.

4.19. Stage 2—linear hardening

In stage 2 there is substantial resistance to dislocation motion which can arise from a number of causes, some of which may be more or less important in different materials. These are as follows:

(1) Since like dislocations on the same plane repel one another (section 4.8, p. 58) there will be a back stress if dislocations at the head of the queue are held up by an obstacle—this could be a tangle of dislocations which cuts the plane or a grain boundary or a precipitate.

(2) A dislocation on one plane has to pass a piled-up group (as in (1)) on a parallel plane (analogous to the interaction of single dislocations in stage 1). This pile-up of N dislocations acts as a super-dislocation of Burgers' vector Nb and it will have correspondingly larger elastic strains and interaction forces.

(3) Dislocations have to cut through 'forest' dislocations which thread the glide plane.

(4) When a dislocation on one plane cuts through another as in (3) one or both of them can be left with a 'jog' on it in which one part of the dislocation has moved relative to the other part by a Burgers' vector (Fig. 4.10). Depending on the actual dislocations which are involved the motion of this 'jogged' part of the dislocation can be nonconservative and so it will lead to the formation of vacancies or interstitials. This has already been discussed in connection with dislocation climb (section 4.6, p. 55). There will be a drag on the dislocation and so the applied stress needed to continue to move it will increase.

All these mechanisms will operate more strongly as the dislocations multiply during further deformation and so the process of work-hardening proceeds.

4.20. Stage 3—cross-slip

In stage 3 the rate of work hardening decreases. This is due to the fact that at a certain high stress a dislocation at the head of a piled-up group which was produced in stage 2(1) can now avoid any obstacle which is blocking its path

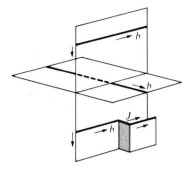

FIG. 4.10. The intersection of dislocations can leave a 'jog' *J* on either or both of them. Depending on the Burgers' vectors of the dislocations, the motion of *J* can be either conservative or non-conservative. For the intersection of two screw dislocations drawn here, *J* will have non-conservative motion as it is moved downwards and this constitutes a drag on the dislocation motion.

by transferring itself to a neighbouring parallel plane. In order to do this it must first travel along one of the intersecting slip planes (Fig. 4.11). This process, which can only be produced by screw dislocations (see section 4.6) is called cross-slip. Once the stress is sufficiently high for a dislocation to avoid one obstacle by cross slip it will be able to avoid other obstacles in a similar manner and hence the hardening rate becomes less rapid. Cross-slip can only occur in crystals such as f.c.c. and b.c.c. which have several equivalent slip planes. It is not observed in h.c.p. crystals.

Nearly all the dislocation interactions which we have discussed can be assisted by thermal activation. Some of the features of the stress–strain curves will depend therefore on the specimen temperature. At liquid-helium temperature the work-hardening rate in stage 1 tends to be increased and the onset of stage 3 is suppressed. At high temperatures stage 2 is shortened and stage 3 begins at a lower applied stress (Fig. 4.12).

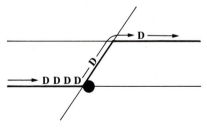

FIG. 4.11. Cross slip. Dislocations piled up against an obstacle give rise to hardening. If the stress is increased sufficiently, screw dislocations can slip via an intersecting plane on to one which is parallel to the main slip plane. Thereafter this process can continue and so the rate of work hardening (in stage 3) is reduced.

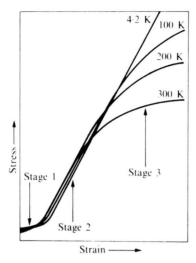

FIG. 4.12. Idealized stress strain curves for a f.c.c. crystal at various temperatures showing how the lack of thermal activation at low temperatures tends to increase the rate of work hardening in stage 1 and inhibits the onset of cross slip in stage 3.

4.21. The yield drop

The yield drop which is observed in b.c.c. crystals (but not the alkali metals) is *not* an intrinsic property of the material; it is in fact dependent on the testing machine. Nearly all deformation experiments are carried out by extending the specimen at a *constant rate* by a 'hard' driving mechanism. If, when the crystal yields, the dislocations multiply very rapidly, then the specimen tries to extend more quickly than the motion of the tensile machine itself.

The yield-drop effect is most pronounced for crystals in which dislocations are tightly locked so that there are only a small number which can move easily at the initial yield point. Since the deformation has to be the same as that imposed by the machine the stress increases sharply giving rise to a very rapid dislocation multiplication and this then leads to a yield drop. Impurity locking of dislocations in b.c.c. crystals is particularly effective and this is why the yield drop is most pronounced in these materials. Eventually the hardening mechanisms which we have already discussed will become operative and the specimen work-hardens in the normal manner.

4.22. The fracture strength

So far we have considered what happens when a crystal is deformed plastically. What determines the point at which the specimen will fracture? The tensile strength of a material is by no means a simple quantity. The conditions for fracture essentially set in when the increased stress produced by the decrease in cross-sectional area as extension proceeds is no longer compensated for by the increase in work-hardening of the material. Since it is in stage 3 that the work-hardening does not increase so rapidly, fracture usually occurs during this part of the stress–strain curve.

4.23. Brittle fracture

The deformation mechanisms which we have described in this chapter give rise to *plastic flow* before ultimate failure of the sample. Many materials, however, such as glass and cast iron, exhibit *brittle fracture* in which there is virtually no plastic deformation before failure occurs.

Brittle fracture is always associated with the existence or the formation of cracks either on the surface of the material or inside it. The stress at the tip of a crack is much higher than the average stress within the material and so there will be a tendency for yielding to occur at the tip in order to relieve the stress. This can be done in two ways: either (a) plastic deformation can occur—this work-hardens the material at the tip of the crack and so further yielding at that point cannot take place or (b) under suitable conditions the crack can open up and run rapidly right through the sample thereby causing brittle fracture without prior deformation or warning.

If the stress for plastic yielding is lower than that for crack propagation then brittle fracture will not occur. But we have already seen (section 4.20) that as the temperature is reduced the yield stress for plastic deformation increases. This is particularly so for the b.c.c. metals. Crack propagation can then become dominant and brittle failure can occur below a certain temperature. Hence b.c.c. metals such as iron and tungsten exhibit brittle failure as the temperature is reduced.

The cracks which initiate brittle fracture can be formed in a number of ways: e.g. during the production of the material, as, for example, on the surface of glass; or they may be produced by small inclusions as, for example, graphite flakes in cast iron; they can occur at weak spots in the intercrystallite boundaries of polycrystals, or they may be formed by dislocation interactions in the crystal.

By reducing the number of these cracks, or by preventing them from propagating, the chance of brittle failure can be reduced. For example, a very fine glass fibre is much stronger than bulk glass because it has a very small surface area and so the chances of it having surface cracks are thereby very much reduced. Cast iron with rounded carbon precipitates is stronger than samples with flaky precipitates because the stresses at their tips are smaller. In general, because grain boundaries are obstacles to crack propagation, materials with very small crystallites, with their larger number of boundaries, are stronger than coarse-grained samples.

Most theories of brittle fracture begin with the assumption that the energy which is put into the system by the external stress must be at least equal to the extra surface energy which is required to form the new faces which are exposed as the crack propagates. This process is discussed in detail in section 15.9.

4.24. Small-angle grain boundaries

Up till now we have shown how dislocations can be used to explain phenomena associated with deformation, but there is a purely geometrical aspect of dislocations which is important.

We have already mentioned (section 2.11, p. 29) that even good single crystals can have a mosaic structure in which small misorientations of the lattice can occur between one part of the crystal and another. In Fig. 4.13(a) we show two parts of a crystal which have slightly different orientations. However, the gaps between the two sections will to the best of their ability be filled by atoms as shown in Fig. 4.13(b). We see that these extra part-planes of atoms which have been inserted form edge dislocations where they stop, i.e. at A, B, and C. Thus a small-angle grain boundary can be considered to be made up of an array of dislocations. If θ is the angle between the crystallites then the spacing between the dislocations will be given by b/θ. Thus if θ is only $1'$ of arc ($\sim 10^{-3}$ rad) this spacing will be $\sim 2.5 \times 10^{-7}$ m. Conversely if there is a dislocation in the crystal, the immediately neighbouring material will have slightly different orientations to the main crystal.

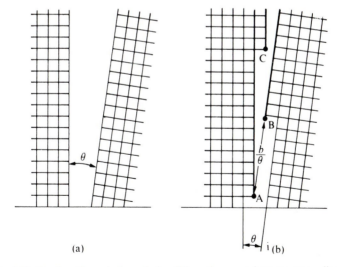

(a)   (b)

FIG. 4.13. Small-angle grain boundaries. When the space between two adjacent crystallites (a) is filled in by atoms the boundary (b) forms an array of edge dislocations.

Similar considerations can be applied to a crystal which has been *plastically* bent to a radius of curvature R (Fig. 4.14). Part planes must be added to the outer part of the crystal to take up the extra strain, otherwise the specimen would spring back to its original shape when it was released.

If s and $s+ds$ are the lengths of the inner and outer arcs respectively and b is the lattice spacing then the number of part planes which are necessary to accommodate ds is ds/b.

But

$$\frac{s+ds}{R+dR} = \frac{s}{R} = \frac{ds}{dR} \tag{4.8}$$

where dR is the specimen thickness. Hence

$$ds = s\,dR/R = A/R, \tag{4.9}$$

where A is the area of the specimen surface. So the total number of dislocations ds/b is from (4.9) equal to A/bR and the *dislocation density* N_d is

$$N_d = 1/(bR). \tag{4.10}$$

Thus a crystal which is slightly bent to a radius of curvature of 1 m will have $N_d \sim 4 \times 10^9 \text{ m}^{-2}$.

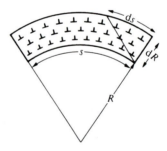

FIG. 4.14. A crystal which is plastically deformed to a radius of curvature R contains dislocations which allow for the extra strain ds on the convex side of the crystal (see text).

4.25. Crystal growth

Another field to which dislocation concepts have contributed is that of crystal growth. When crystals are grown from the liquid or vapour it is well known that it is difficult for very small crystallites to be nucleated because their vapour pressure is so high that they re-evaporate; e.g. stable water droplets have to be formed on dust particles which act as nuclei. The initial deposition of large nuclei would require a highly supersaturated solution.

If we already· have a seed crystal in the solution atoms are first deposited at the angles and corners on its surface because they are more tightly bound at these places. Eventually, however, in the idealized situation shown in Fig. 4.15, all the corners will be used up and fresh nuclei must be formed on a smooth surface for which a highly supersaturated solution would be necessary, otherwise the rate of crystallization would decrease.

In practice it is found that crystals can grow from solutions which are only slightly supersaturated, and they continue to grow at the same rate. It would therefore appear that the corners, shown in Fig. 4.15, are *not* used up. Frank proposed that growth occurs where screw dislocations emerge from a surface. We have already seen (Fig. 3.14), that these transform the lattice into a spiral ramp and it is clear that if atoms are deposited on the step of the ramp it rotates about the central screw dislocation, but it never vanishes.

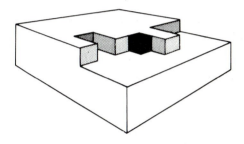

FIG. 4.15. Crystal growth on a surface of a crystal which does *not* contain screw dislocations would result in all the corners and steps being preferentially filled by atoms. This would then leave a perfectly smooth surface on which it would be very difficult to produce more nuclei for further crystallization.

This model of crystal growth has been verified by many remarkable photographs in which growth spirals and terraces are clearly visible on the surface of crystals. One such example is shown in Fig. 4.16.

If two screw dislocations of opposite sign emerge from the surface a series of terraces is produced, rather than a spiral ramp.

4.26. The direct detection of dislocations

Many ingenious techniques have been developed to detect the presence of dislocations in crystals. Of these, by far the most important has been the direct observation by electron microscopy of dislocations in thin metal films.

Electron microscopy

Whilst it is now possible in some electron microscopes under favourable conditions, to resolve individual planes of atoms, such high resolution

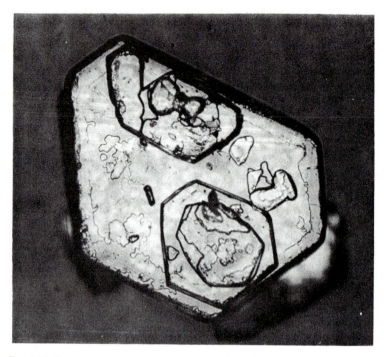

FIG. 4.16. Growth spirals and terraces on the surface of a cadmium iodide crystal. (Magnification × 260.)

(~ 2 to 3×10^{-10} m) is not required to detect dislocations. Instead advantage is taken of the fact, already discussed in section 4.24, that in the neighbourhood of a dislocation the crystal orientation changes very slightly.

When a single crystal is placed in the electron microscope the electrons which pass through the specimen are diffracted into angles which are determined by the Bragg formula (2.1). For any arbitrary angle of incidence the image which is produced will be completely featureless, but if the specimen is rotated so that a strong diffracted beam passes through the elements of the microscope, then the slight orientation changes around the dislocations give a sharp change in image contrast and the position of the dislocations can clearly be seen. The appearance of the dislocations can be made to vary as the crystal is rotated, but under the best conditions they show up as sharp dark lines (Fig. 4.17). It should be appreciated that these lines are *not* direct images of the dislocations but they are a diffraction pattern which is due to their presence. The lines are much thicker ($\sim 100 \times 10^{-10}$ m) than are the dislocations themselves.

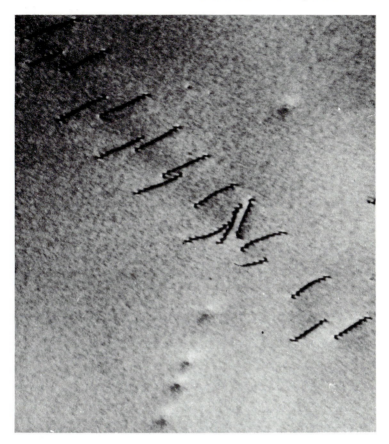

FIG. 4.17. A transmission electron micrograph of dislocations in a thin specimen of Cu:10 % Al. Two rows of dislocations on parallel planes (top left to bottom right) are clearly imaged. Each dislocation terminates at the upper and lower surfaces of the specimen. The line of dots (nearly vertical) is another row of dislocations whose orientation is such that they are not imaged so satisfactorily. They are seen to interact with the first set of dislocations at the centre of the picture. (Magnification × 50 000.) (Photo by courtesy of Dr. H. P. Karnthaler.)

Electrons are absorbed very strongly in crystals and so in conventional electron microscopes (accelerating voltage 50–100 kV) the specimens have to be extremely thin—about 0·1 μm—and this is achieved by careful chemical or electro-chemical dissolution. In 1 MV microscopes specimens about 1 μm thick can be used. In all cases the image on the screen is a projection through the entire thickness of the specimen, so that the ends of the dislocations in Fig. 4.17 are actually terminating at the top and bottom surfaces of the specimen.

In favourable situations the heat from the electron beam is sufficient to cause slight stresses in the specimen and the motion and interaction of the dislocations can be observed. It is also possible to directly deform the specimen within the microscope and to observe the dislocation reactions, although such a manipulation of small specimens within the electron microscope is a difficult and tricky technique.

4.27. Other methods for the detection of dislocations

For completeness we mention the basic principles which are involved in some other detection techniques.

(1) Electron microscopy of crystals of large organic molecules which contain heavy atoms (e.g. Pt phthalocyanine) in which the heavy-atom planes are $\sim 12 \times 10^{-10}$ m apart. These planes can be resolved quite easily and the configuration of the atomic planes in edge dislocations can be determined. This technique was very important in the original detection of dislocations by electron microscopy, but its application is limited to very special materials.

(2) X-ray topography. The change in orientation around the dislocation which is used in electron microscopy can also be used in the technique of X-ray topography; changes in the intensity of the diffracted beam as it passes through different parts of the specimen give an indication of the position of the dislocations. Since however we still have no good X-ray lenses, the magnification is limited to that which may be obtained by direct optical magnification of the photographic image. It is, however, very useful for examining highly perfect crystals which have a low dislocation density.

(3) The change in orientation around the dislocation also produces a broadening of the X-ray lines in a conventional diffraction experiment. This has been a very useful method for determining the dislocation density, particularly in work-hardened specimens which contain many dislocations.

(4) Etch pits. The places where dislocations end at the surface of a crystal are more (or less) prone to chemical attack than the rest of the surface. By very careful chemical or electro-chemical etching the ends of the dislocations can be shown up as etch pits (or hills) on the specimen surface (Fig. 4.18).

(5) Field-ion microscopy techniques (section 7.13, p. 120), in which the position of individual atoms at the tip of a fine needle can be observed, are being developed to investigate the detailed atomic arrangement at a dislocation.

(6) Decoration of dislocation networks. Because impurities segregate preferentially around dislocations (section 4.5, p. 54) they may be 'decorated' in certain cases by special techniques. In transparent crystals which contain silver, the silver atoms around the dislocations can be 'developed out' by the

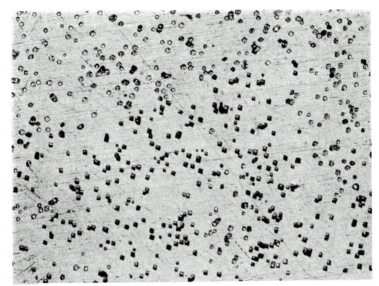

FIG. 4.18. Etch pits on the surface of aluminium show where the dislocations terminate. Note that some etch pits are square while others are hexagonal. These probably correspond to dislocations of differing Burgers' vectors. (Magnification × 210.)

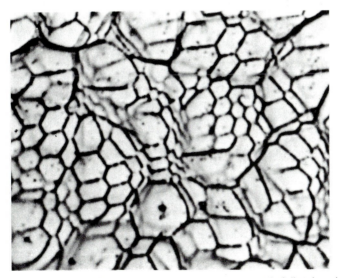

FIG. 4.19. A decorated dislocation network in a transparent crystal of silver bromide. (Magnification × 2875.) (Photo by courtesy of Professor J. W. Mitchell.)

photographic effect and the patterns of dislocation networks can be studied in detail (Fig. 4.19). It is interesting to note that the first 'direct' observation of dislocations was made by this technique (Hedges and Mitchell 1953). Even this list is not exhaustive, but it shows the very large range of artillery that has been brought to bear on the problem of the direct observation of dislocations in crystals and which has been so very successful in translating the concept of dislocations from being just a good idea to becoming of immense importance in our understanding of crystal behaviour.

PROBLEMS

4.1. A copper crystal has a homogeneous dislocation density of 10^{10} m^{-2}. Estimate the elastic strain energy of a screw dislocation per metre. (For Cu, shear modulus $= 4 \times 10^{10}$ N m^{-2}. Burgers' vector $= 0.26$ nm.) Compare your answer with that for Problem 3.9 (p. 52).

4.2. Show that if the line tension of a dislocation is T then the force per unit length of dislocation which is necessary in order to bend the dislocation to a radius of curvature R is T/R.

An aluminium crystal contains imperfections which act as pinning points for dislocations. If the imperfections are homogeneously distributed, with a density 10^{24} m^{-3}, estimate (a) the force per unit length of dislocation which will make it bow out into a semicircle, and (b) the shear stress which must be applied to the crystal in order to achieve this effect. (For Al, shear modulus $= 2.6 \times 10^{10}$ N m^{-2}, Burgers' vector $= 0.3$ nm.)

4.3. The yielding of a particular crystal is dominated by the stress which is required to operate a Frank–Read source. If the yield stress is 5×10^8 N m^{-2}, the shear modulus is 3×10^{10} N m^{-2} and the Burgers' vector is 0.3 nm, calculate the minimum length of the operative Frank–Read sources.

4.4. Estimate the yield strength of a copper–silica alloy which has spherical precipitates of silica, each of radius 100 nm. The volume fraction of silica is 1 per cent, the shear modulus is 4.2×10^{10} N m^{-2} and the Burgers' vector is 0.26 nm.

4.5. When one dislocation cuts through another a jog, equal in length to a Burgers' vector is formed on one dislocation, thereby making it longer (Fig. 4.10). Estimate (a) the energy of a jog in aluminium, (b) the force per unit length of dislocation for one dislocation to cut through a homogeneous dislocation 'forest' of density 10^{18} m^{-2} which is aligned perpendicular to the first dislocation, and (c) the external shear stress necessary for (b).

4.6. The elastic energy which is released when a crack of unit width opens up is of the order of $L^2\sigma^2/Y$, where L is the depth of the crack, σ is the tensile stress around the crack, and Y is Young's modulus. This energy is used to create the fresh surfaces of the crack. If E_s is the surface energy per unit area show that the critical stress which is necessary for a crack to propagate is given by

$$\sigma_{crit} = (E_s Y/L)^{\frac{1}{2}}.$$

Estimate the minimum crack depth which would initiate brittle fracture for a material with $Y = 6 \times 10^{10}$ N m^{-2}, $\sigma_{crit} = 5 \times 10^8$ N m^{-2} and $E_s = 10^2$ J m^{-2}.

4.7. An aluminium crystal is bent to a radius of curvature of 5 cm. What is the minimum dislocation density in the material (Burgers' vector = 0·3 nm)?

4.8. A narrow X-ray beam 2×10^{-5} m in diameter is used to obtain a diffraction pattern from a metal crystal. From the broadening of the spots it can be deduced that over the area covered by the beam there is a change in lattice orientation of 3 minutes of arc. Assuming that the dislocations are homogeneously distributed estimate the minimum dislocation density in the crystal if the Burgers' vector is 0·25 nm.

4.9. Calculate the dislocation spacing in a symmetric 1° tilt boundary in a copper single crystal (Burgers' vector = 0·26 nm).

5. The thermal vibrations of the crystal lattice

5.1. Thermal energy and lattice vibrations

OUR general ideas about temperature and thermal equilibrium are based on the fact that the individual particles in a system are endowed with some type of vibrational motion which increases as the temperature is raised. In a solid the energy associated with this vibration and perhaps also with the rotation of the atoms or molecules is called the *thermal energy*. In a gas the translational motion of the atoms and molecules will also contribute to this energy.

A full appreciation of the thermal energy is fundamental to an understanding of many of the basic properties of solids. For example, we would like to know the value of this energy and how much is available to scatter a conduction electron in a metal, since this scattering gives rise to an electrical resistance. Or the energy might be used to activate a crystallographic or a magnetic transition, or a dislocation interaction. We are also interested to know how the vibrational energy changes with temperature because this gives a measure of the heat energy which is necessary in order to raise the temperature of the material. (It will be recalled that the specific heat is the thermal energy which is required in order to raise the temperature of unit mass, or of 1 g mole, by one kelvin.)

For simplicity we shall only discuss the thermal energy which is due to the vibration of atoms in a solid. It is usual to discuss this in terms of the classical theory of atomic vibrations and then to show that this is completely unsatisfactory at low temperatures. Quantum mechanics is then brought in with a round of applause to save the day. Whilst this approach was intriguing during the first half of this century it would now appear to be a little dated. We therefore intend to give a unified discussion.

5.2. Statistics

The system of vibrating atoms in a crystal is of course very complicated and a calculation of the total thermal energy from a knowledge of the energy of each individual atom is clearly impossible. However, if the system is in thermal equilibrium (i.e. if it is all at the same temperature T) then we do have rules which give us the *relative probability* that a particle in that system will have an energy, say E_1, rather than E_2. This probability function $f(E)$ is usually referred to as the *statistics* of the system. Depending on the type of system and its constituent particles, there are three possible functions which may be used, and we quote these without proof.† For completeness we shall

† Derivations and discussion can be found in any book on statistical mechanics, e.g. D. K. C. MacDonald (1963), *Introductory statistical mechanics for physicists*, Wiley, New York; J. S. Dugdale (1966), *Entropy and low temperature physics*, Hutchinson, London.

discuss all three functions in this section, although we shall only use the Maxwell–Boltzmann statistics in this chapter.

In all the systems which we shall consider it is assumed that there are no interactions between the particles, i.e. the energy state of one particle is not influenced by that of its neighbour. Systems may be divided into two main types depending on whether they are comprised of particles which are *distinguishable* or *indistinguishable*. Distinguishability in this context is a rather formal concept and it is used to describe whether *in principle* the particles can be distinguished from one another. For our purposes the only system of distinguishable particles is the assembly of atoms that forms a solid. In such a system each atom can be designated by a unique set of position coordinates. Indistinguishable particles, e.g. an atom in a gas or an electron in a metal, are best thought of as being described by a wave packet which is sufficiently extended in space that it overlaps the wave packets of the other particles. If this occurs it is impossible even in principle to distinguish one particle from another.

Systems of distinguishable particles are described by the Maxwell–Boltzmann statistics

$$f_{MB} = A \exp(-E/kT). \tag{5.1}$$

where k is Boltzmann's constant.

Indistinguishable particles must be divided into two types. If they have half-integral spin (e.g. electrons) and hence are subject to the Pauli exclusion principle they obey the Fermi–Dirac statistics

$$f_{FD} = [\exp\{(E-E_F)/kT\} + 1]^{-1}. \tag{5.2}$$

where E_F is a parameter called the Fermi energy.

If the particles have zero or integral spin (e.g. photons, phonons, or ^{4}He nuclei) they are controlled by the Bose–Einstein statistics

$$f_{BE} = [\exp\{(E-\alpha)/kT\} - 1]^{-1} \tag{5.3}$$

It should be noted that in the limit of high energies both f_{FD} and f_{BE} approach f_{MB} and this approximation is often made to simplify calculations.

The quantities A, E_F, and α are normalizing parameters, i.e. their value is adjusted so that when each function is summed over all the energy states which are available to a particle in that system the result is equal to the total number of particles. When the function f_{BE} is applied to a system of photons or phonons, where the number of 'particles' is not constant with time, the parameter α is set equal to zero.

5.3. Density of states

In the previous section we placed no restrictions on the energy which a particle may have. However, quantum mechanics tells us that if a particle is constrained in any way—whether it be an electron bound to an atom or an apple in a box—or in our case an atom bound in a crystal—then the energy of the particle can only have certain special discrete values. It cannot increase infinitely gradually from one value to another—it has to go up in steps. Depending on the system we are considering these steps can be so small that the graduation from one permitted energy to the next can be assumed to be continuous; this is the case for the apple in a box, and it is the realm of classical mechanics. But on an atomic scale we very often have to take account of the fact that the energy can only jump by a discrete amount from one value to another.

As we shall see later in this chapter each particular energy level may be associated with more than one different *state* (or wave function) of the system. If this occurs the level is said to be *degenerate*. If a level is, say, doubly degenerate the chances are twice as great that a particle will have that energy than if it were not degenerate. More generally, if $g'(E)$ is the degeneracy of the level associated with the energy E, then the probability that a particle will have that energy will be

$$g'(E)f(E). \tag{5.4}$$

The normalization of the function f can be expressed by

$$\sum_E g'(E)f(E) = 1, \tag{5.5}$$

where the summation is over all permitted values of E. Sometimes, instead of normalizing using (5.5), it is more convenient to express the relative probability as

$$g'(E)f(E) \bigg/ \left\{ \sum_E g'(E)f(E) \right\}. \tag{5.6}$$

$f(E)$ does not then need to be normalized and so, for example, A in (5.1) can then be set equal to unity.

The mean energy of a particle will be given by

$$\sum_E E g'(E)f(E) \bigg/ \left\{ \sum_E g'(E)f(E) \right\}. \tag{5.7}$$

If the energy levels are very close together their distribution may be treated as a continuous function of E and the terms in (5.7) may be written as integrals instead of summations. We must then replace the degeneracy $g'(E)$ by the *density-of-states* function $g(E)$, where $g(E)$ is the number of discrete states per unit energy interval, and so the number of states between E and $E + dE$ is $g(E)\,dE$. The number of *occupied* states in that interval is of the form

$g(E)f(E)\,dE$ and the mean energy will be

$$\int_0^x Eg(E)f(E)\,dE \bigg/ \int_0^x g(E)f(E)\,dE. \qquad (5.8)$$

In the present chapter the density of states will be used to describe the number of atomic oscillators per unit energy (or frequency or wave vector) range.

5.4. The mean energy of the harmonic oscillator

We assume that the energy of the atoms in a crystal is governed by the Maxwell–Boltzmann statistics and so from (5.1) we see that as the temperature is raised the probability that an atom has a higher energy is increased. We therefore have the general picture that the energy of the atomic vibrations becomes greater as we go from low temperatures to higher ones. These changes in energy, however, are discontinuous. A quantum-mechanical treatment of the one-dimensional simple harmonic oscillator (and we assume that in one dimension the vibrating atom approximates to this) shows that its permitted energies are $\tfrac{1}{2}\hbar\omega$, $(1+\tfrac{1}{2})\hbar\omega$, $(2+\tfrac{1}{2})\hbar\omega$,....$(n+\tfrac{1}{2})\hbar\omega$, where ω is the angular frequency,† n is an integer and $\hbar$ is Planck's constant$/2\pi$. The step between one permitted energy and the next is therefore $\hbar\omega$. It should be noted that the minimum energy of the system is not zero but $\tfrac{1}{2}\hbar\omega$. This will be present even at 0 K, and it is called the zero-point energy. Thermal energies are most conveniently measured from this baseline as $\hbar\omega$, $2\hbar\omega$,..., $n\hbar\omega$. For the one-dimensional oscillator each of these energy levels is non-degenerate and so using (5.1) and setting $g'(E) = $ unity in (5.7) the mean energy of the oscillator is given by

$$\tfrac{1}{2}\hbar\omega + \sum_n n\hbar\omega \exp\left\{\frac{-n\hbar\omega}{(kT)}\right\} \bigg/ \sum_n \exp\left\{\frac{-n\hbar\omega}{(kT)}\right\}. \qquad (5.9)$$

The summations in (5.9) can be performed by a neat trick. First substitute $x = \hbar\omega/kT$ in the exponents and we have

$$\hbar\omega \sum_n n \exp(-nx) \bigg/ \sum_n \exp(-nx);$$

this may be written as

$$-\hbar\omega\frac{d}{dx}\left\{\log \sum_n \exp(-nx)\right\}.$$

The summation is a straightforward geometrical progression giving

$$-\hbar\omega\frac{d}{dx}\left[\log\left\{\frac{1}{1-\exp(-x)}\right\}\right] = \hbar\omega\frac{\exp(-x)}{1-\exp(-x)} = \frac{\hbar\omega}{\exp(x)-1}.$$

† Angular frequency $\omega = 2\pi \times$ frequency.

The mean energy is therefore

$$\tfrac{1}{2}\hbar\omega + \frac{\hbar\omega}{\exp{(\hbar\omega/kT)} - 1}. \qquad (5.10)$$

This energy can be considered either as the time-averaged energy for a particular atom, or it can be thought of as the average energy of *all* the atoms in the assembly at any instant in time. At low temperatures, i.e. when $\hbar\omega > kT$, the exponential in the denominator of (5.10) is large and the mean energy will be just slightly higher than $\tfrac{1}{2}\hbar\omega$. At high temperatures, i.e. $\hbar\omega < kT$ the exponential can be expanded as $1 + \hbar\omega/(kT)$ and so the mean energy is $\tfrac{1}{2}\hbar\omega + kT$ or approximately kT. The mean energy is then independent of the frequency of oscillation. This is the *classical limit* because the energy steps, $\hbar\omega$, are now small compared with the mean energy of the oscillator.

5.5. Specific heat at room temperature

Since an atom can vibrate independently in three dimensions it should really be considered to be equivalent to three separate oscillators and its energy will be three times that given by (5.10) (assuming that ω is constant). The total *thermal* energy for the system of N atoms (ignoring the $\tfrac{1}{2}\hbar\omega$ term) at high temperatures will then be $3NkT$. The specific heat, i.e. the heat required to change the temperature by 1 degree, will therefore be $3Nk$ or $3R$, where R ($= Nk$) is the gas constant. This is the *Dulong–Petit law*—the specific heat of a given number of atoms of any solid is independent of temperature and is the same for all materials. Usually the specific heat per gram atom is quoted so that N is Avagadro's number (6.02×10^{23}).[†] The value of the specific heat is then $24.9 \ \mathrm{J} \ K^{-1} \ (\mathrm{g \ atom})^{-1}$.

5.6. Specific heat at low temperatures

As the temperature is reduced the specific heat remains constant at $3Nk$ until T is so small that $\hbar\omega \sim kT$. We then have to use the full expression (5.10) for the thermal energy. The specific heat will be given by differentiating (5.10) with respect to T. This gives an expression (first suggested by Einstein in 1907) which is dominated by a term which decreases exponentially as the temperature is reduced. Thus the effect of the quantum theory is to predict a specific heat which *decreases* as the temperature is reduced and this is in general accord with experiment. This fall-off from the value of $3Nk$ will begin at a temperature of the order of $\hbar\omega/k$. The average value for the frequency of vibration of atoms in many solids is about 10^{13} Hz and so the temperature below which the specific heat tends to fall comes out to be about 400 K.

---

† This is for a monatomic crystal. For compounds the specific heat is often quoted for a quantity of material which contains Avagadro's number of *atoms*; e.g. for NaCl it would be quoted for *half* a gram mole.

Thus for some materials the specific heat can be less than the Dulong–Petit value even at room temperature and certainly as the temperature is reduced below room temperature the specific heat will decrease.

It is important to remark that the *exponential* decrease in the specific heat, mentioned in the last paragraph, is not observed. The fall-off is actually not quite so rapid, and at low temperatures (usually in the liquid-helium region) the specific heat is found to vary as T^3. The reason for this is that the Einstein treatment described above assumes some kind of an average value ω for the atomic vibrational frequency, and this turns out to be an over-simplification. Clearly to describe the complicated pattern of atomic vibration in a crystal will require a wide spectrum of frequencies. In particular, any low-frequency oscillators which are present will have small values of $\hbar\omega$, and hence their thermal energy will remain at the value kT (and the specific heat at k) to much lower temperatures than will the energy associated with oscillators of higher frequency. So as the temperature is reduced the lower frequency vibrations will contribute more than their fair share to the specific heat and this modifies the exponential fall and makes it less drastic.

5.7. The Debye theory of the specific heat

A detailed calculation on these lines would require a knowledge of how many atoms were vibrating with a frequency ω_1, how many with ω_2, and so on; or in general terms we would need to know the *frequency spectrum* of the atomic vibrations. We can write this formally as a density of states $g(\omega)$ (section 5.3), where $g(\omega)\,\mathrm{d}\omega$ represents the number of oscillators lying within the frequency range ω to $\omega + \mathrm{d}\omega$. The energy for each oscillator is given by (5.10) and hence for the assembly of atoms the mean thermal energy (neglecting the $\frac{1}{2}\hbar\omega$ term) would then be

$$\langle E_{\mathrm{th}} \rangle = \int [g(\omega)\hbar\omega \, \mathrm{d}\omega / \{\exp (\hbar\omega/kT) - 1\}] \tag{5.11}$$

where the integration is taken over the whole range of atomic vibrational frequencies.† The specific heat is obtained, as before, by differentiating (5.11) with respect to T.

In order to go further there are two problems which must be solved. First, what is the actual form of the function $g(\omega)$, and second, what is the range of frequencies over which the function in (5.11) should be integrated?

---

† We assume that the distribution of oscillators is quasi-continuous in ω, and so we may use integration instead of summation. This could not be done in the derivation of the mean energy of a single oscillator (5.9) where the individual quantum steps $\hbar\omega$ might be large compared with kT.

In principle the job can be done properly. If we have a linear chain of particles joined to each other by springs, then we can calculate the possible modes of vibration of that system. The crystal can be considered to be a three-dimensional array of linked particles, and if we know the dependence of the interatomic forces on the atomic spacing and the crystal structure then we can calculate the modes of vibration of the atoms in the crystal. Computers have made these calculations less tedious than they were, but this approach does not lend itself to any simple analysis.

Nevertheless, we do have a clue here as to how to proceed. We must take account of the fact that the atoms are coupled to one another and we have to consider the vibrations as being a property of the assembly of atoms as a whole rather than as the motion of isolated atoms. Clearly, if the atoms were really isolated they could not achieve thermal equilibrium nor could the crystal conduct heat, and so the interaction approach is a reasonable one; but looking down on all the vibrating atoms as from a bug's eye view, we see a tremendously complicated pattern of motion. How can we simplify it? We have already dealt with this in section 2.6 (p. 24) when we discussed Fourier analysis. A complicated pattern can be built up by the superposition of certain simple wave patterns.

This is the basis of the Debye theory of specific heats (1912). It is assumed that hypothetical oscillators generate simple sine waves throughout the crystal and these will displace the atoms away from their equilibrium positions by an amount equal to the amplitude of the sine wave at that point. If we have a whole set of such oscillators generating sine waves of certain frequencies and amplitudes then we might hope that the superposition of such waves will simulate the complicated pattern of the actual atomic vibrations.

5.8. Standing waves in a crystal

There are two sets of waves which can be selected, either of which can be used to solve the problem.† These are standing waves and travelling waves. In order to be able to calculate the complete set of either of these two types of wave we need to define some *boundary conditions* for the problem. For standing waves (and we shall concentrate on these in our treatment) it is mathematically convenient that they have nodes (i.e. zero amplitude) at the crystal boundaries. They are waves of the same type as those which can be sustained on a stretched string of length L (Fig. 5.1). Their permitted wavelengths will therefore be $2L/n$ (or wave number $\pi n/L$), where n is any positive integer.

---

† This calculation is a very important one, not merely in the present context but also for the case of electrons in metals (Chapter 7) in which we shall merely quote the results which are derived here.

FIG. 5.1. Boundary conditions for standing waves on a stretched string. The permitted modes of vibration are those in which there is a node at each end of the string.

In three dimensions it is much more convenient to work in terms of the wave vector k because as we have already discussed (section 2.4, p. 21) this can be resolved into three components k_x, k_y, k_z, along the three main axes of the sample. By analogy with the one-dimensional case, the permitted values of the three components of k for standing waves in a crystal which has sides L_x, L_y, L_z will be $\pi n_x/L_x$, $\pi n_y/L_y$, $\pi n_z/L_z$. The resultant wave vector k will be given by

$$k^2 = k_x^2 + k_y^2 + k_z^2 = \pi^2(n_x^2 + n_y^2 + n_z^2)/L^2 \tag{5.12}$$

$$= \pi^2 n^2/L^2 \tag{5.13}$$

where $n^2 = n_x^2 + n_y^2 + n_z^2$ and we assume $L_x = L_y = L_z \equiv L$.

Note that for a given magnitude of k there will be several combinations of integers n_x, n_y, n_z.† Each different set will correspond to a wave vector with a different *direction*, i.e. it relates to a different hypothetical oscillator. We need to calculate the number of oscillators for a given value of k but since the n are integers it is clear from (5.12) that not all values of k are permitted. (It should be noted that this is another example of discrete levels or states which occur in a bounded system.) We therefore cannot calculate the number of oscillators for an arbitrary value of k because there may not be any—we can only calculate the number whose wave vectors lie within a certain range of k-values, say from k to $k + dk$.

5.9. The oscillator distribution

A simple way of doing this can be deduced from (5.12). It is clear that, if the components n_x, n_y, n_z for a given k are plotted in three dimensions using ordinary orthogonal axes, then the resultant vectors n will lie on the surface of a sphere (actually only an octant of a sphere because the n are here restricted to being positive). This is shown in Fig. 5.2. A possible combination of the n within the sphere occurs at each crossing point within the figure. Each corresponds to a permitted value of k and hence to an oscillator which produces that k. On average there is one crossing point to each elementary cube, and since the n are integers the volume of each cube is unity. Thus if the radius of the octant is n, the number of different states possible up to that value of

† This is an example of degeneracy (section 5.3).

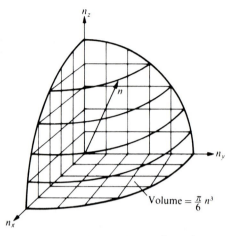

Fɪɢ. 5.2. An octant of a sphere drawn with coordinates in n-space. Each crossing point for integral values of n_x, n_y, n_z, corresponds to a permitted mode of vibration. There is thus one unit cube in n-space per mode. All modes lying on a spherical surface of radius n will (in an isotropic model) have the same frequency and energy.

n is just the number of elementary unit cubes in the volume, i.e. it is equal to the volume itself, $\pi n^3/6$. From (5.13) we now express n in terms of k:

$$n = V^{\frac{1}{3}}k/\pi, \tag{5.14}$$

where we have written V, the volume of the crystal, for L^3 and so the total number of oscillators for wave vectors from zero up to k is

$$\pi n^3/6 = Vk^3/(6\pi^2) \equiv G(k). \tag{5.15}$$

The number of oscillators per unit range of k is obtained by differentiating $G(k)$ with respect to k. Writing g for dG/dk, we then obtain

$$g(k)\,dk = Vk^2\,dk/(2\pi^2), \tag{5.16}$$

where $g(k)$ is the density of states as a function of the wave vector (section 5.3, p. 80).

The formulae (5.15) and (5.16) are very important because they relate to the number of oscillators or stationary states of waves which can be maintained in a specimen of volume V. This is independent of the type of wave we are considering. In the present case we are interested in the waves which represent the displacements of the atoms from their equilibrium positions but the same ideas and formulae will be used when we consider the matter waves associated with electrons in our study of the properties of metals.

If we refer to (5.11) we see that because all the original calculations were made in terms of the frequency we really need $g(\omega)$ rather than $g(k)$. If we assume that $k = \omega/v$, where v is the velocity of the wave, a simple substitution in (5.15) gives

$$G(\omega) = V\omega^3/(6\pi^2 v^3), \tag{5.17}$$

and so

$$g(\omega)\,d\omega = V\omega^2\,d\omega/(2\pi^2 v^3). \tag{5.18}$$

It is important to note that the density of states varies as the *square* of the frequency.

5.10. Polarization

Whilst we have now obtained an expression for the density of states of the waves in terms of their frequency, there is one aspect which we have yet to consider. This is their polarization. The waves produce atomic displacements which can be resolved so as to be parallel to the wave vector (longitudinal waves) and in two directions at right angles to it (transverse waves). Thus each state in (5.18) really corresponds to three independent waves of different polarizations—two transverse and one longitudinal†—and so the actual density of states is three times that given by (5.18). This is analogous to our treatment of the Dulong–Petit law (section 5.5).

5.11. Debye theory; the thermal energy

The mean thermal energy of the system, given by (5.11), may now be determined by substituting for $g(\omega)$ and inserting the factor of 3 for the polarizations. We obtain

$$\langle E_{\text{th}} \rangle = \frac{3V}{2\pi^2 v^3} \int \frac{\hbar\omega^3}{\exp(\hbar\omega/kT)-1}\,d\omega. \tag{5.19}$$

We are now nearly home, and the only problem which remains is to select the range of frequencies over which (5.19) should be integrated. We should perhaps here run over the argument we have been using. We have assumed that the atomic displacements can be considered to be produced by the superposition of standing waves which affect all the atoms and which are generated by hypothetical oscillators. These have a frequency distribution given by the density of states $g(\omega)$, (5.18), a continuous function. We then used the mean thermal energy of a real vibrating system (5.10) as the energy of each of these oscillators. We have previously suggested that the number of independent oscillators in the system is equal to three times the total number of atoms, i.e. $3N$, and so we now argue that, independent of the model that is used to

† It should be noted that the assumption that the polarizations are mutually perpendicular and strictly longitudinal and transverse is *not* generally true for a discrete lattice, except in simple symmetry directions.

analyse the atomic motion, the total number of oscillators which actually produce the displacements must also be $3N$, because we cannot change the number of degrees of freedom of the system.

We therefore assume that all permitted frequencies can be excited from zero up to a frequency ω_{max}, where ω_{max} is the frequency corresponding to a total of $3N$ oscillators being operative. We can use (5.17) to obtain a relation between $3N$ and ω_{max} (remembering an extra factor of 3 for the different polarizations).

$$3N = V\omega_{max}^3/(2\pi^2 v^3). \tag{5.20}$$

Now substitute for V/v^3 in (5.19) to give

$$\langle E_{th} \rangle = \frac{9N}{\omega_{max}^3} \int_0^{\omega_{max}} \frac{\hbar\omega^3}{\exp(\hbar\omega/kT) - 1} \, d\omega. \tag{5.21}$$

This expression can be simplified by introducing a temperature θ, which is known as the Debye characteristic temperature and which is defined by

$$k\theta = \hbar\omega_{max}, \tag{5.22}$$

so that θ is the temperature at which the mean classical thermal energy of an oscillator is equal to a single quantum step $\hbar\omega_{max}$.

On substituting $x = \hbar\omega/kT$ and $x_{max} = \theta/T$ into (5.21)

$$\langle E_{th} \rangle = (9NkT^4/\theta^3) \int_0^{x_{max}} \frac{x^3}{e^x - 1} \, dx. \tag{5.23}$$

The specific heat, c, is most easily obtained by differentiating (5.21) with respect to T and then, as above, changing the variable to x. Hence

$$c = 9R(T/\theta)^3 \int_0^{x_{max}} \frac{e^x x^4}{(e^x - 1)^2} \, dx, \tag{5.24}$$

where we have taken 1 mole of material and have written $Nk = R$.

5.12. The specific heat at high and at low temperatures

Eqn (5.24) is the Debye specific heat function; it is plotted in Fig. 5.3. At high temperatures, when x is small, the integral in (5.23) reduces to $x^2 \, dx$ and hence the energy becomes $3RT$ and on differentiation this yields the Dulong–Petit value of $3R$ for the specific heat. At low temperatures, where x is large, we can approximate by allowing the upper limit in (5.23) to go to infinity. The integral then has the value of $\pi^4/15$. On differentiating, we obtain for the specific heat

$$c = 1941(T/\theta)^3 \text{ J mol}^{-1} \text{ K}^{-1}. \tag{5.25}$$

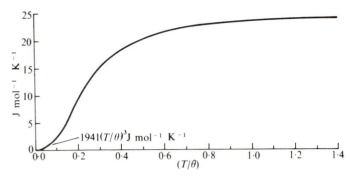

FIG. 5.3. The specific heat per mole according to the Debye theory, calculated from (5.24). Below $T/\theta = 0.1$ the curve follows the cubic relation indicated in the figure.

Thus at low temperatures the specific heat should decrease as T^3. (It will be recalled (section 5.6, p. 82) that the Einstein single-frequency theory predicted an exponential dependence.) A rough calculation shows that this cubic dependence should begin at temperatures below $\theta/10$ and, although in practice it usually sets in at a lower temperature than this, the T^3 variation of the specific heat at very low temperatures is well established experimentally. It should be noted that the decrease in the specific heat at low temperatures is very substantial. A simple calculation using (5.25) shows that if θ is 300–400 K, the specific heat at 4.2 K is about $10^{-4} \times$ the room-temperature value.

The success with which (5.24) represents the specific heats of most solids is truly remarkable when one considers the simple terms to which the complicated vibrations of the crystal lattice have been reduced. The only adjustable parameter which is used is θ, and this it will be recalled (5.22) is a measure of the maximum frequency of atomic vibrations.

5.13. The Debye θ

It is interesting to discuss the significance of the parameter θ because it is constantly referred to in the literature, not merely in connection with specific heats but with many other thermal properties as well. Since θ is proportional to the maximum vibrational frequency of the atoms then, by analogy with a simple vibrating system, a high value of the frequency, and hence of θ, implies that we are dealing with a lattice which has very strong interatomic forces and light atoms. Thus diamond has a value of θ of about 2000 K. Conversely, lead, which has weakly bound heavy atoms, has a low value of θ—about 100 K. Most common elements of medium atomic weight have a θ of a few hundred kelvins, e.g. copper is 348 K and aluminium is 426 K. θ can be considered as the temperature at which all the oscillators with a frequency ω_{max} are excited. Since (5.18) shows that the density of oscillators is proportional to ω^2, this means that at θ K we can assume that most of the oscillators have a frequency fairly close to ω_{max}.

5.14. The minimum wavelength

It is instructive to calculate the wavelength of the vibrations at ω_{max}. We assume that the displacements are propagated through the crystal with the velocity of sound, which in copper is about 4000 m s^{-1}. From (5.20) ω_{max} is $348k/\hbar$, and so the wavelength $= 2\pi v/\omega = 4000h/348k = 5.4 \times 10^{-10}$ m, which is about twice the atomic spacing. This is an example of a general rule that at θ the wavelength of the lattice vibrations is about twice the lattice constant (i.e. the wave vector is half a basic vector of the reciprocal lattice). This result is important because since the real crystal consists of a pattern of discrete particles we could not ascribe a unique interpretation to a vibration which had a very short wavelength. The motion of the particles under these conditions could then be described just as well by a vibration with a longer wavelength (Fig. 5.4). In fact, provided the wavelength is less than twice the atomic spacing it can always be reduced to a wave with a longer wavelength. It should be emphasized that the waves we have been discussing only have a real physical significance at the atomic positions. What the wave does at other places may be of mathematical interest, but it has no bearing at all on the behaviour of the system. The mathematical treatment of waves on a linear chain of identical atoms is dealt with in Problem 5.10.

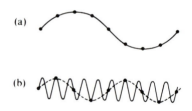

Fig. 5.4. (a) The atomic displacement for lattice vibrations of long wavelength. (b) The atomic displacement when the wavelength is short compared with the atomic spacing. In this case the motion could be described just as well by the dotted line which has a wavelength longer than the atomic spacing.

5.15. Phonons

Our analysis of the atomic vibrations can be looked at in a slightly different way. We can consider the crystal lattice as one large system which is able to pulsate at various frequencies like a lump of jelly. Previously we have suggested that these vibrations were generated by hypothetical oscillators, but of course it is the system as a whole that is generating its own oscillations—there is really no need to invent an external driving mechanism. The super-position of all the calculated displacements at the various points within the system must be the same as the atomic displacements at those points. So whilst at the beginning of this chapter we discussed the vibration of isolated atoms, we now see that because they are coupled together, a more logical point of view is one in which we are thinking in terms of the various modes of vibration of a single vibrating system—the whole assembly. But even this composite system must still obey the rules of quantum mechanics. Its energy can only increase in discrete steps of $\hbar\omega$, only now these quanta of vibrational energy involve displacements of *all* the atoms rather than just one of them. By analogy with photons (the quanta of electromagnetic radiation) these quanta of lattice vibrational energy $\hbar\omega$ are called *phonons*. If the temperature is raised, the amplitude of the atomic vibrations increases, and in quantum terms this can be thought of as being due to an increase in the number of phonons in the system.

It is interesting to note that phonons, which are associated with the entire lattice, are obviously 'particles' which are indistinguishable. They should therefore be controlled by the Bose–Einstein statistics, since they have zero spin (section 5.2). The probability that a state of energy E is occupied is then given by (5.3)

$$f_{BE}(E) = \{\exp(E/kT) - 1\}^{-1}.$$

The mean energy of the state associated with a frequency ω and an energy $\hbar\omega$ is therefore the energy $\times$ the probability of occupation, which is

$$\frac{\hbar\omega}{\exp(\hbar\omega/kT) - 1}.$$

This is exactly the same as the temperature-dependent part of the expression (5.10) which we used as the mean energy of a *localized* atomic oscillator. There is therefore no inconsistency in treating all the vibrations of the individual atoms as if they are a superposition of vibrational quanta (phonons) which are associated with the entire assembly of atoms.

5.16. Travelling waves; cyclic boundary conditions

The idea that the atomic displacements can be represented by a set of standing waves is usually satisfactory if the whole crystal is in thermal equilibrium. Nevertheless it is clearly rather restrictive because it precludes any possibility of a disturbance at one end of the crystal being propagated to the other end as, for example, in the flow of heat. For this type of situation a travelling-wave solution would appear to be more realistic. The problem in choosing a set of travelling waves is to define the boundary conditions satisfactorily.

For a one-dimensional linear chain of length L this can be done easily by adopting a simple device. We assume that the chain is joined together at its ends and that the wave continually travels around the loop (Fig. 5.5). It is clear that the amplitude of the wave at a point x must be the same as that at $(x + L)$ because these are actually the same point. If we represent the wave† as $\sin (k_x x - \omega t)$ then we must have

$$\sin (k_x x - \omega t) = \sin \{k_x(x + L_x) - \omega t\} \quad \text{for all } x \text{ and } t. \quad (5.26)$$

Thus

$$k_x L_x = 2\pi n_x$$

or

$$k_x = 2\pi n_x / L_x. \quad (5.27)$$

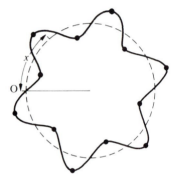

Fig. 5.5. Cyclic boundary conditions for a chain. The position of an atom on the chain as measured from O is x. The chain is assumed to be bent round and joined at its ends. If its entire length is L then the atomic displacement at x must be the same as that at $x + L$ since these are actually the same point.

This condition (5.27) is a factor of 2 greater than the permitted values of k_x for standing waves (5.12), but this is discounted by the fact that for standing waves we only used positive values of k_x, whereas for travelling waves both positive and negative values of k_x are needed to take account of waves travelling in either direction. However, a standing wave can be considered to be formed by the superposition of two travelling waves of equal but opposite wave vectors, and so the standing wave treatment implicitly includes negative wave vectors, but they always occur paired with the positive value and they

† We write in the time dependence here to emphasize the fact that it is a travelling wave, but it plays no part in the analysis and it can be omitted.

are not treated separately in the analysis. On adapting our previous analysis to travelling waves, therefore, we must sum over a whole sphere of radius n and not just an octant (5.15).

The extension of these cyclic boundary conditions to three dimensions is mathematically quite straightforward, although it requires a considerable stretch of the imagination to visualize the actual situation. It is quite easy to consider the lattice being made continuous by bending the crystal round so that two surfaces touch thereby forming a torus, but to visualize doing this in three dimensions is tortuous. It is easier to consider the imposition of cyclic boundary conditions in three dimensions by imagining that we have an infinitely large crystal in which the vibrational pattern repeats itself in three dimensions at intervals of L. Mathematically the justification for the cyclic boundary conditions is that they give the correct number of degrees of freedom ($3N$) for the system.

The extension of (5.27) to three dimensions gives an expression analogous to (5.12) for k^2:

$$k^2 = k_x^2 + k_y^2 + k_z^2 = 4\pi^2(n_x^2 + n_y^2 + n_z^2)/L^2, \qquad (5.28)$$

and the extra factor of four in (5.28) (and hence eight in k^3) just counters the fact that we must calculate the volume of the entire sphere of radius n instead of the octant of the sphere.

PROBLEMS

5.1. A crystal is composed of identical atoms which only interact with one another very weakly. Each has a set of three equally spaced singlet energy levels, E_1, E_2, E_3, such that $E_2 - E_1 = 10^{-23}$ J and $E_3 - E_1 = 2 \times 10^{-23}$ J. Calculate the temperature at which (a) there are equal numbers (to within 1 per cent) of atoms in each of the three states. (b) 99 per cent of the atoms are in the ground state E_1. (c) If the energy level E_2 is doubly degenerate calculate the temperature for (b). (d) Why in (a) cannot truly equal populations be achieved?

5.2. In an assembly of 10^{23} simple harmonic oscillators each has a frequency 10^{13} Hz. Calculate (ignoring the zero-point energy) the mean thermal energy of the system at 2 K, 20 K, 200 K, and 2000 K. Discuss the significance of your answers. In what temperature range could the system be treated classically?

5.3. Using the Dulong–Petit relation calculate the (classical) thermal energy of 1 g mole of a material at 300 K. Aluminium has a Debye θ of about 430 K. Estimate the thermal energy per gram mole at 300 K with the aid of Fig. 5.3. Explain why the two results are different.

5.4. What factors determine the Debye θ of a particular element? Estimate the Debye θ of gold from the following information. For gold, the atomic weight is 197, the density is 1.9×10^4 kg m^{-3} and the velocity of sound is 2100 m s^{-1}. For copper the atomic weight is 63.5, the density is 8900 kg m^{-3}, the velocity of sound is 3800 m s^{-1}, and the Debye θ is 348 K. Copper has the same structure as gold.

5.5. At low temperatures the Debye θ of NaCl and KCl. which have the same crystal structure, are 310 K and 230 K respectively. The lattice specific heat of KCl at 5 K is 3.8×10^{-2} J mol^{-1} K^{-1}. Estimate the lattice specific heat of NaCl at 5 K and of KCl at 2 K.

5.6. Estimate the change in the thermal energy of 100 g of copper (atomic weight = 63.5, Debye θ = 348 K) when it is cooled (a) from 300 to 4 K, (b) from 78 to 4 K, and (c) from 20 to 4 K. If the latent heat of liquid helium is 2700 J l^{-1}, estimate how much helium would be required to produce the cooling in each case. (This calculation will over-estimate the quantity because the cooling power of the cold gas is neglected.)

5.7. The Debye θ of diamond is 2000 K, its density is 3500 kg m^{-3}, its atomic weight is 12 and the interatomic spacing is 0.15 nm. Calculate the velocity of sound in diamond. Make an estimate of the dominant phonon wavelength at 300 K (see section 6.4) and calculate the frequency of lattice vibration to which this corresponds.

5.8. Write a computer programme to evaluate the Debye integral (5.25) as a function of T/θ. At what value of T/θ does the expression deviate from T^3 behaviour by 5 per cent?

5.9 Compute the value of the temperature-dependent part of the mean energy of a simple harmonic oscillator (5.10) and plot a curve to show how this energy, in units of $\hbar\omega$, varies with T. At what value of kT (in the same units) is the mean energy (a) $0.1 \hbar\omega$, (b) $0.5 \hbar\omega$, (c) $\hbar\omega$?

5.10 Derive the dispersion relation, i.e. the ω versus k dependence, for a linear chain of similar atoms of mass m, spaced a distance a apart. Assume that only nearest neighbour interactions are important and that the restoring force on each atom is proportional to its relative displacement with respect to its neighbours (force constant = K). Proceed as follows: let the displacements of the $(n-1)$, n, and $(n+1)$ atoms be u_{n-1}, u_n, and u_{n+1}, and calculate the relative displacements of the nth atom with respect to its neighbours. Equate the restoring force to $m\,d^2u_n/dt^2$. Assume a solution of the form $u_n = C \exp i(\omega t - kna)$ and hence show that $\omega = \sqrt{K/m} |\sin \frac{1}{2}ka|$. Note that this relation repeats when k reaches multiples of $\pm \pi/a$.

6. Phonons in non-metals; thermal conductivity

IN the previous chapter we introduced the concept of the phonon as being a quantum of lattice vibrational energy. This arose from the treatment of a system of standing or travelling waves in a box which, when superimposed, simulate to some degree the complex pattern of the atomic vibrations in a crystal. If the whole crystal is at a uniform temperature the standing wave model is quite acceptable and we do not need to know precisely where the phonons *are*, because they are uniformly distributed.

6.1. The effect of a temperature gradient

If, however, we set up a temperature gradient then clearly the amplitude of the atomic vibrations at the hot end of the specimen will be greater than at the cold end. The standing wave model is now unsatisfactory because this cannot produce the different amplitudes of atomic vibration at various parts of the specimen, nor would it be able to deal with the fact that heat will now flow from a hotter region to a colder one.

The mathematical problem (which we shall not pursue) when there is a temperature gradient is by no means trivial, but an extension of the idea of travelling waves (section 5.16) leads to the concept of wave *packets* (a super-position of travelling waves) which can travel within the crystal transferring the energy of thermal motion from one region to another. These elementary wave packets are again called phonons. They will be more dense at the hotter end of the crystal than at the colder end and we can think of the process of heat conduction as being due to the diffusion of phonons down a concentration gradient to the cold end of the specimen.

This duality between phonons as non-localized standing waves and as localized, particle-like wave packets is of course entirely analogous to the non-localized photon in a black-body enclosure and the corpuscular, wave-packet photon which excites a photo-electric cell.

6.2. Thermal conductivity

In a heat-conduction determination we often use a rod-shaped specimen, one end of which is maintained at a constant temperature, whilst the other end is heated by a known amount of electrical power $\dot{Q}$. After a time dynamic equilibrium is established and the temperature gradient $\Delta T/\Delta x$ along the specimen is measured by two thermometers spaced Δx apart. The thermal conductivity κ is then defined by

$$\kappa = \dot{Q}/(A \, \Delta T/\Delta x),$$

where A is the cross-sectional area of the specimen.

In order to develop a theory of thermal conductivity we need to know the rate at which phonons can transfer heat along a crystal. These phonons are being continually scattered by various barriers or interactions which we shall discuss later, and we may assume that they travel some average distance l, the *mean free path*, between these collisions.

It turns out that we can adapt the classical kinetic theory of the heat conductivity of gases† to the problem of heat transport by the phonon quasiparticles and we may use the kinetic formula

$$\kappa = \tfrac{1}{3}c_v l v, \tag{6.1}$$

where c_v is the specific heat per unit volume‡ of the phonons and v is their velocity. (Since phonons are elastic waves within the crystal, v is effectively the velocity of sound.)

We have already discussed the behaviour of the specific heat in Chapter 5, and we know that it is constant at high temperatures and it varies as T^3 at low temperatures; v is almost constant. Our main problem now is to determine the behaviour of l, the mean free path.

6.3. Scattering mechanisms for phonons

There are several scattering mechanisms which can operate to limit the value of l. We shall not deal with them all in detail, but they include:

    (a) the interaction of phonons with one another ('*umklapp*-processes');
    (b) scattering by point defects (impurities, isotopes, etc.);
    (c) scattering by the boundaries of the specimen or crystallites;
    (d) scattering by dislocations.

Each mechanism will have an associated mean free path l_a, l_b, l_c,... for which there will be thermal resistances W_a, W_b, W_c,.... These act as resistances in series so that the overall thermal resistance W is

$$W = W_a + W_b + W_c + \cdots .$$

The mean free paths must be combined as reciprocals to give an overall effective value of l,

$$1/l = 1/l_a + 1/l_b + 1/l_c + \cdots .$$

If all the interactions were equally strong the situation would be very complicated and we shall limit our discussion to the cases where only one particular scattering mechanism is important.

---

† For example, see Sir James Jeans (1940), *An introduction to the kinetic theory of gases*, Chapter 2, Cambridge University Press.

‡ It should be particularly noted that c_v is the specific heat per unit volume and *not* the specific heat at constant volume.

6.4. Dominant wavelength of phonons

In considering the interactions which can produce phonon scattering and hence give rise to a thermal resistivity it should always be borne in mind that whilst one might tend to visualize phonons as particles which dart around within the crystal, they do have a wave-like nature. The strength of any interaction will in general depend on the wavelength.

As the temperature is changed the phonon spectrum will alter. At high temperatures, as we have seen in section 5.13 (p. 90), waves with a frequency near to ω_{max} dominate. This corresponds to a wavelength which is of the order of twice the atomic spacing $2a$. At low temperatures only long-wavelength phonons are excited. A rough rule which gives the order of magnitude of the dominant phonon wavelength λ_d (wave vector k_d) at a low temperature T is

$$\lambda_d \approx (\theta/T)a, \quad \text{or} \quad k_d \approx (T/\theta)2\pi/a \tag{6.2}$$

where θ is the Debye characteristic temperature.

Thus for many materials λ_d is a few hundred atomic spacings in the liquid helium region, whereas it is about $2a$ at room temperature.

6.5. The interactions of phonons with one another

Everyday observation gives the impression that there is little or no interaction when groups of waves cross each other's path. Two sets of ripples on a pond pass through one another and one set is not scattered by the other. This is because these waves are propagated in a *linear* medium, i.e. one in which the displacement at any point is proportional to the applied force. The waves in such a medium are *harmonic* and the energy of the wave is proportional to the square of its amplitude. The resultant displacement of the medium at any point can be obtained by adding the separate amplitudes of the individual waves at that point.

If, however, the medium is not exactly linear, so that the expression for the displacement also depends on quadratic and possibly higher powers of the force, then the principle of superposition breaks down. The waves are said to be *anharmonic*. If two waves of angular frequencies ω_1 and ω_2 are propagated through the medium the output will also contain waves of frequency $\omega_1 + \omega_2$ and $\omega_1 - \omega_2$ (and possibly others). This is a *non-linear* system and a well known example is the mixer stage of a superheterodyne radio.†

A medium can usually be treated as linear if the amplitude of the waves is small, because then any higher-order (anharmonic) terms may be neglected. For phonons in crystals this is the case at very low temperatures and then the interaction of phonons with one another is negligible. At higher temperatures,

---

† Note that the phenomenon of 'beats' in acoustics which is produced by the superposition of two neighbouring frequencies is *not* a non-linear effect.

when the atomic displacements are larger, the anharmonic terms become appreciable and so two waves can interact and some portion of their energy can be converted into a wave with a different wave vector and frequency. The smallest quantum which can be 'removed' from each wave to form the third is a phonon, and this process is usually described by saying that two phonons (one from each of the two individual waves) have combined to form the resultant third phonon. The major portions of the two waves, however, actually passed on without interaction and these may be ignored in any discussion on scattering processes.

The conservation laws for the combination of phonons are:

(1) energy is conserved, i.e. $\hbar\omega_1 + \hbar\omega_2 = \hbar\omega_3$.   (6.3)

(2) wave vector is conserved,† i.e. $k_1 + k_2 = k_3$.   (6.4)

How does this type of phonon interaction contribute to the thermal resistance? The answer is that in many circumstances it does not. If in Fig. 6.1(a) we are considering the heat transport along the specimen in a horizontal direction, then the vector addition of two phonons with wave vectors k_1

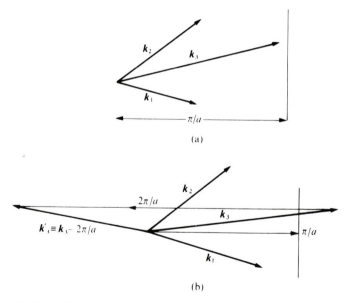

(a)

(b)

Fig. 6.1. The addition of two wave vectors. $k_1 + k_2$ to form k_3. (a) Normal process. (b) Umklapp process; if k_3 extends beyond the boundary π/a then it is physically equivalent to the vector k_3' which differs from k_3 by $2\pi/a$.

† The conservation of wave vector (whether for electrons or phonons) is only a stringent requirement if the structure possesses *translational symmetry*, i.e. if it is crystalline. It does not necessarily have to hold in a glass or if a defect is involved.

and k_2 yields a resultant k_3 in which the energy transport along the specimen is unchanged after the interaction. This is called a *normal process* or *n-process*. n-processes play an important part in establishing thermal equilibrium, but they do not directly contribute to heat conduction.

6.6. Umklapp processes

If, however, either or both the incident phonons have fairly large wave vectors the resultant k_3 in (6.4) may be so large that the corresponding wavelength is *shorter* than twice the interatomic spacing, i.e. $k_3 > \pi/a$. We have already described in section 5.14 (p. 90) how in a *discrete* lattice such a disturbance is completely equivalent to a wave with a longer wavelength. It may be shown that in one dimension this equivalent wave has a wave vector which is equal to $k_3 - 2\pi/a$.† Thus if k_3 is greater than π/a (and less than $2\pi/a$) the equivalent wave vector will be *negative* and so the phonon velocity is *reversed* (Fig. 6.1(b)).‡ This means that if the sum of the two oncoming phonons is large enough, their resultant will be a phonon with the same total energy, but which is travelling in the *opposite* direction. This represents a diminution in the heat transport down the temperature gradient, i.e. it produces a thermal resistance.

This type of mechanism is called an *umklapp§* process, or u-process. It gives rise to considerable confusion, not merely because of its name, but also because the idea that two waves can combine to form another one which is travelling backwards is outside our normal experience. It should be emphasized, however, that this only occurs because the medium is *discrete* and *periodic*. In a continuous medium u-processes cannot occur.

The u-process can be understood by referring to Fig. 6.2. The upper section shows the atomic displacements at a time t which are described by the wave indicated by the heavy line and which has a wavelength which is *shorter* than $2a$. The same displacements may also be described by the wave drawn as a dashed line and this has a wavelength *greater* than $2a$. Fig. 6.2(b) shows the situation a short time later when the heavy curve has travelled to the *right*. It can be seen that the new atomic displacement can also be described by the equivalent, longer wavelength (dashed curve) which has moved to the *left*. Thus the displacements of the atoms can be described just as well by a wave moving in one direction as by another wave moving in the opposite direction. The figure should make it quite evident that u-processes, which depend on the complete physical equivalence of the two waves, can only occur in a discrete lattice.

† In three dimensions the equivalent wave has a wave vector which is equal to $k_3 \pm G$, where G is a basic vector of the reciprocal lattice (section 2.7. p. 26).

‡ Note that the direction of k gives the direction of the *phase velocity* of the wave. Complete equivalence of the two waves implies that the direction of energy flow is the same for each and this is given by the *group velocity*. This is in the same direction as k for $|k| < \pi/a$ but it is in the opposite direction of k for $|k| > \pi/a$. Thus when $|k|$ becomes greater than π/a the direction of energy flow is reversed. See problem 6.6

§ From the German *umklappen*, meaning to flop over.

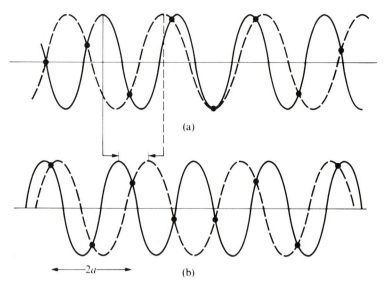

FIG. 6.2. Umklapp process. The dots represent atomic positions at some instant (a). Their displacement from their mean position can be described by the solid line which has a wavelength slightly shorter than $2a$ or, equivalently, by the dotted line which has a wavelength slightly longer than $2a$. If we assume that, as indicated by the arrow, the solid wave moves to the right, (b), to give the new atomic displacements, then the dotted (longer) wave has to move to the left to give the same displacements. Thus these two waves, which are physically equivalent, move in opposite directions.

6.7. The probability for the occurrence of a u-process

If two phonons combine to yield a third one which has $k \geqslant \pi/a$ it is necessary that at least one of the initial ones has $k \geqslant \pi/2a$. Since this is half the maximum permitted wave vector, such a phonon will have about half the maximum permitted energy, which from the Debye theory will be about $\frac{1}{2}k\theta$ (section 5.13, p. 89). The probability that such a phonon will be excited is (ignoring the Bose–Einstein statistics!) proportional to $\exp(-k\theta/2kT)$, or $\exp(-\theta/2T)$.

Thus this type of interaction increases exponentially with the temperature. At high temperatures however, when $T \gtrsim \theta$, nearly all the phonons will have a large enough k to produce u-processes, and the probability will then be proportional to the total number of phonons which are present. At high temperatures[†] this is just proportional to T.

[†] At high temperatures the thermal energy $3NkT \approx nh\omega_{max}$, where ω_{max} is the maximum vibrational angular frequency of the lattice (section 5.13). Thus the number of phonons n is proportional to T.

The thermal resistance W_u due to u-processes therefore first shows an exponential increase as the temperature is raised, and it then slows down to a T-dependence.

6.8. The scattering of phonons by point defects

The scattering of phonons by obstacles depends on the size of the obstacle relative to that of the phonon wavelength. The smaller the obstacle size the less is the scattering, because the wave can sweep around it. Point defects (and here we must also include the strain field around the defect) are of atomic dimensions. At low temperatures the dominant phonon wavelength is very much larger (section 6.4, p. 97) than the defect and under these conditions we may use the same theory as that developed for the scattering of light by very small particles (Rayleigh scattering). This gives a scattering probability which is proportional to k^4, i.e. to T^4 using (6.2) and this implies a mean free path which varies as T^{-4}. If this is inserted into the kinetic theory equation (6.1) then over the low-temperature range, where the specific heat is proportional to T^3, the effect of impurity scattering will be to give a thermal conductivity which varies as T^{-1}, i.e. a thermal resistance W_{imp} which is proportional to T. A more precise calculation suggests that W_{imp} varies as $T^{\frac{1}{2}}$.

At higher temperatures, where the phonon wavelength is comparable with the defect dimensions and where the specific heat is constant, W_{imp} becomes independent of temperature.

6.9. Isotopes

When we talk of an impurity we mean any point defect which upsets the perfect regularity of the crystal lattice. High purity materials can still be composed of elements containing mixtures of isotopes of different masses which will be randomly distributed throughout the specimen. The lattice therefore will not be as regular as that for an isotopically pure sample. The minority isotope acts as if it were an impurity. The effect of this may be demonstrated very convincingly. The heat conductivity of an isotopically pure crystal is indeed higher than that of a specimen containing a mixture of isotopes (Fig. 6.3).

6.10. Boundary scattering

Both u-processes and impurity scattering, which have been described in the previous sections, decrease as the temperature is reduced. The mean free path of the phonons therefore increases and, indeed, it may become as large as the diameter of the specimen itself. Once this has occurred it can clearly increase no further. The phonon mean free path becomes constant and the conductivity is then limited by diffuse scattering of the phonons at the sample boundaries. (If the boundaries are very well polished, specular reflection can occur and this will increase the mean free path.)

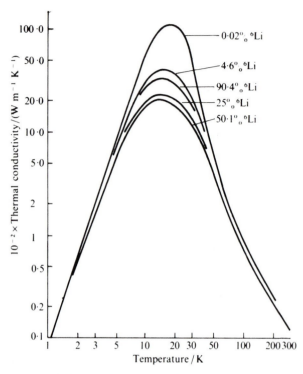

FIG. 6.3. The isotope effect in the thermal conductivity of LiF. The bottom curve is for a specimen in which the lithium content is almost exactly 50 per cent ^{6}Li and ^{7}Li. As the isotopic purity is increased in either the ^{6}Li or the ^{7}Li direction the thermal conductivity increases. (After R. Berman and J. C. F. Brock (1965). *Proc. Roy. Soc.* A **289**, 46.)

Using the kinetic theory (6.1) with a constant value of l equal to the smallest dimension d of the specimen we obtain a thermal conductivity which varies as the specific heat (i.e. as T^3 at low temperatures) and which is also proportional to d. Thus the boundary thermal resistance W_{bound}, varies as $(dT^3)^{-1}$. This is well borne out by experiment (Fig. 6.4).

In a polycrystalline sample the value of d will be restricted by the size of the crystallites and not by the specimen diameter.

6.11. Dislocation scattering

We discuss the effect of dislocations only briefly because no theory has yet been developed which accounts entirely satisfactorily for the magnitude of the effect. We have already seen in section 4.7 (p. 56) that around a dislocation

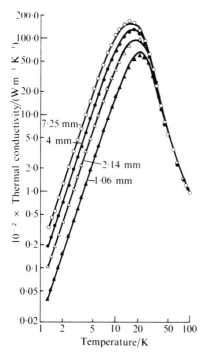

FIG. 6.4. The boundary scattering of phonons and its effect on the thermal conductivity of LiF. As the width of the specimen is decreased the thermal conductivity in the low temperature region is reduced. (After P. D. Thacher (1967). *Phys. Rev.* **156**, 975.)

there is a long-range strain field. This will scatter phonons and it gives rise to a thermal resistance W_{dis}, which is proportional to the dislocation density and which varies as T^{-2}. This type of temperature-dependence has been observed in crystals in which the dislocation density has been increased by deformation.

6.12. A summary of the behaviour of phonon conductivity

We shall now assemble the results of the preceding sections. In an ideal pure crystal the thermal resistance over most of the temperature range will be determined by u-processes. At low temperatures, however, owing to the exponential decrease in the probability of these processes, the mean free path of the phonons will increase until it becomes limited by the dimensions of the specimen, whereafter it must remain constant (Fig. 6.5(a)). The phonons will then be scattered by the boundaries of the specimen, and at sufficiently low temperatures this gives a conductivity which is proportional both to T^3 and to the size of the specimen.

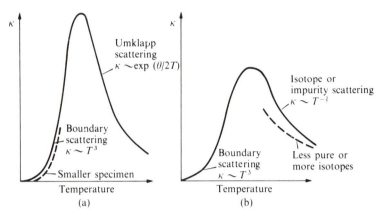

FIG. 6.5. The general behaviour of the phonon thermal conductivity in a non-metal at low temperatures for (a) a pure specimen in which u-processes dominate the rise in conductivity and (b) in an impure specimen (isotopically or otherwise) in which phonon scattering by impurities is dominant.

Point defects and mixtures of isotopes will upset this behaviour. The exponential increase in the conductivity due to the u-processes will be suppressed in favour of a conductivity which is proportional to $T^{-\frac{3}{2}}$ (Fig. 6.5(b)), although at sufficiently low temperatures boundary scattering will again become dominant. If the specimen is polycrystalline, the size of the crystallites will determine the conductivity.

Large amounts of impurity and dislocations will decrease the conductivity. The dislocations will be effective at low temperatures, whilst the effect of the impurities is more important at higher temperatures.

6.13. The absolute magnitude of the thermal conductivity

Non-metals are generally considered to be poor conductors of heat and, indeed, in many cases they are. Under certain circumstances, however, they can be very good conductors. If the material is a highly perfect crystal and is sufficiently pure so that W_u dominates at high temperatures, we see from section 6.7 (p. 100) that this will be smaller for a high value of θ. The most outstanding example is diamond, for which θ is about 2000 K. At room temperature the thermal conductivity of a good diamond can be as high as 2000 W m^{-1} K^{-1}, whereas that of copper is 400 W m^{-1} K^{-1}. The high conductivity of diamond combined with the fact that it is an electrical insulator has been utilized in the design of some thin-film semiconducting devices which are kept cool in operation by being evaporated on to a diamond slice.

The peak of the conductivity curve in Fig. 6.5(a), which usually occurs around $\theta/20$, can be very high indeed—of the order of 2×10^4 W m^{-1} K^{-1} in good pure crystals.

The heat conductivity in the boundary scattering region may be calculated very easily using (6.1). Thus, if we have a specimen which has a θ of 300 K, a density of 10^4 kg m^{-3}, and an atomic weight of 100, then c_i at 4 K will be, using (5.25) 460 J K^{-1} m^{-3}. If the velocity of sound is 5000 m s^{-1} and the specimen is 3 mm in diameter, the thermal conductivity at 4 K will be

$$\tfrac{1}{3} \times 460 \times 5000 \times 3 \times 10^{-3} = 2300 \text{ W m}^{-1} \text{ K}^{-1}.$$

6.14. The thermal conductivity of glasses and plastics

The irregular structure of glass and plastics gives rise to a large amount of phonon scattering. The mean free path will be very small and hence the conductivity will be very low—it is of the order of 10^{-1} W m^{-1} K^{-1} at room temperature. Nevertheless the conductivity, especially at low temperatures, shows some very interesting features and these are discussed in section 15.28 *et seq.*

PROBLEMS

6.1. Estimate the mean free path of phonons in germanium at 300 K using the following information. Thermal conductivity = 80 W m^{-1} K^{-1}. Debye θ = 360 K, atomic weight = 72·6, density = 5500 kg m^{-3}, mean velocity of sound = 4500 m s^{-1}. (Assume that all the heat transport is by phonons.)

6.2. The Debye θ of diamond is 2000 K. Calculate the ratio of the thermal conductivity at 50 K to that at 4 K, assuming that boundary scattering is dominant in both cases.

6.3. In a simple cubic lattice of spacing 0·2 nm a phonon travelling in the [100] direction with wave vector $(2\pi/\lambda)$ $1·3 \times 10^{10}$ m^{-1} interacts and combines with another phonon of the same magnitude which is travelling in the [110] direction. Draw a diagram to show the magnitude and direction of the resultant phonon and discuss the significance of the result.

6.4. Estimate the relative importance of u-processes to the thermal resistivity at 100 K and 20 K for a crystal which has a Debye θ of 300 K.

6.5. Calculate the thermal conductivity at 1 K of a rod of synthetic corundum (Al_2O_3) crystal of molecular weight 102 if its diameter is 3 mm. The velocity of sound = 5000 m s^{-1}, the density = 4000 kg m^{-3} and the Debye θ = 1000 K (n.b. eqn (5.25) is the expression for the specific heat of a quantity of material containing Avagadro's number of *atoms*).

6.6. The phonon dispersion relation in one dimension may be written in the form $\omega = C|\sin(ak/2)|$. Sketch the form of the curve for the range $-2\pi/a \leqslant k \leqslant 2\pi/a$. If the velocity of sound in the material is 5000 m s^{-1} and a is 0.3 nm what is the value of C? What are the phase and group velocities for $k = +\pi/2a$, $-\pi/2a$, $+\pi/a$, $+3\pi/2a$? Which of these values of k correspond to equivalent phonons?

7. Free electrons in crystals

7.1. Introduction

IN the previous chapters we have ignored all effects which are produced by the electrons in the crystal. We assumed that the electrons surrounding the atomic nuclei form closed shells and that all are tightly bound to their atoms. This is certainly a valid assumption for materials which are electrical insulators, but in order to understand the properties of metals and semiconductors it is essential to discuss the behaviour of electrons in a crystal lattice.

At the turn of the century, when all physics seemed to be understood, it was recognized that electrons wandering around inside metals were the cause of electrical conductivity. The Drude–Lorenz theory was developed and this assumed that electrons were classical particles, each having a kinetic energy of $\frac{3}{2}kT$. The theory gave an explanation for both the electrical (σ) and the thermal (κ) conductivities and also for the relationship between σ and κ. This is enshrined in the Wiedemann–Franz law which states that at a given temperature T,

$$\kappa/(\sigma T) = L. \quad \text{the Lorenz constant.}$$

The main snag in the theory was that if each electron had an energy of $\frac{3}{2}kT$ then this should have given an extra contribution of $\frac{3}{2}k$ to the specific heat for each electron. This is not observed. The molar specific heat of metals at sufficiently high temperatures is $3R$, just as it is for non-metals (section 5.5, p. 82). It therefore appeared that the electronic specific heat, for some unaccountable reason, was very small.

It was not until quantum mechanics was applied to the electron system by Sommerfeld that the problem was solved. Further developments of the theory now enable us to have a really full understanding of the basic properties not only of metals, but also of the nature of semiconductors and insulators.

7.2. The electronic structure of atoms

We give here a very brief outline of the way electrons are arranged around the atomic nucleus.† We have already described that when a particle is constrained to a particular region quantum mechanics shows that it may only have an energy selected from a discrete set of levels. Since electrons are constrained by electrostatic attraction to the region around the positively charged atomic nucleus they will have such discrete permitted energies. These energies

---

† See also section 11.4. p. 183.

are determined by the 'state' of the electron and this is usually designated by a wave function ψ. The various functions ψ correspond to the different probability distributions of the electron cloud which can occur around the nucleus (they are sometimes loosely referred to as electron orbits). A particular energy may correspond to more than one independent function ψ, and this is called degeneracy (section 5.3, p. 80).

Because electrons have half-integral spin the states which they occupy are controlled by the Pauli exclusion principle which only permits *one* electron to be in any one particular energy state ψ. So long as there is degeneracy, however, this does *not* mean that there is only one electron with a particular value of the energy.

Starting from the lowest energy the electron states for one atom are grouped in special sets of 2, 8, 18, 32.... If all the states in a set are occupied by electrons the completed set is called a *closed shell* and it is very tightly bound to the nucleus. The electron configuration around the nucleus therefore consists of a number of closed shells outside which are the remaining few electrons. These outer electrons are not so strongly bound (Fig. 7.1(a))—one reason for this is that the positive charge which attracts them to the nucleus will be reduced by the screening effect of the inner closed shells. It is these outer electrons which are responsible for the electrical properties.

In a metal such as copper the mean 'radius' of the charge cloud of outer electrons is about 10^{-10} m, and since the atoms are only 2.5×10^{-10} m apart the clouds from neighbouring atoms almost overlap (Fig. 7.1(b)). It is therefore quite easy for an electron on one atom to transfer itself to its neighbour, and so on through the crystal.

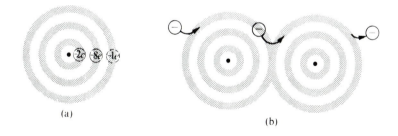

(a)

(b)

Fig. 7.1. (a) A typical configuration of electrons for an atom of a metal, showing the closed (complete) inner shells and the outer shell which usually contains 1, 2, or 3 electrons. The electron trajectories are not necessarily circular. (b) Two neighbouring atoms in a metal crystal are sufficiently close together that electrons in the outer shell can easily pass from one atom to the next.

7.3. The free-electron theory

To a first approximation we can assume that an outer atomic electron can move quite freely within the material and we ignore the fact that when an atom has 'lost' an outer electron then that atom itself becomes a positively charged ion.† This is the idea underlying the 'electron-in-a-box' model.

We therefore need to calculate the possible states ψ and energies E of an electron in a box, where the 'box' is the size of the crystal. The Schrödinger wave equation is

$$-\left(\frac{\hbar^2}{2m}\right)\left(\frac{\partial^2\psi}{\partial x^2}+\frac{\partial^2\psi}{\partial y^2}+\frac{\partial^2\psi}{\partial z^2}\right) = E\psi. \tag{7.1}$$

Since we assume that the electron cannot escape from the box the boundary conditions are that ψ should be zero at all its faces. The solution to (7.1) is then a standing wave‡ which can be written in the form

$$\psi = A \sin k_x x \sin k_y y \sin k_z z. \tag{7.2}$$

The problem is exactly the same as that which we have already encountered for lattice waves in our treatment of the Debye theory of specific heats. The permitted wave vectors are given by (5.12),

$$k^2 = k_x^2 + k_y^2 + k_z^2 = \pi^2(n_x^2 + n_y^2 + n_z^2)/L^2. \tag{7.3}$$

The relation between the energy and k may be obtained by substituting (7.2) in (7.1),

$$\frac{\hbar^2(k_x^2 + k_y^2 + k_z^2)\psi}{2m} = E\psi,$$

and so

$$E = \hbar^2 k^2/2m. \tag{7.4}$$

In the free electron wave function ψ the permitted values of the wave vector k (7.3) are analogous to the quantum numbers n, l, m that characterize the permitted states of an electron which is bound to an atom.

---

† Or we can assume that the ions plus all the other 'free' electrons are smeared out to give a constant background potential, which may also be ignored.

‡ Once again, as in section 5.16 (p. 91), we could have used cyclic boundary conditions and then the solutions of (7.1) would be travelling waves of the form $\psi = \exp(i\mathbf{k}\cdot\mathbf{r})$, where $\mathbf{k}$ can be positive or negative. The final results are again exactly the same as for the standing-wave solution.

Since E is entirely kinetic energy $\frac{1}{2}mv^2$ we may, using (7.4), equate

$$\frac{1}{2}mv^2 = \hbar^2 k^2/2m,$$

and so the momentum is

$$mv = \hbar k.$$

Thus k can be considered to be a measure of the momentum (and hence of the velocity) of the *free* electron. However, in the band theory (Chapter 8) we shall see that this is not always the case, and for this reason k is often referred to as the quasi-momentum.

There is one slight complication that we must bear in mind in calculating the number of permitted states for electrons. Apart from the value of the wave vector (which in the free-electron model gives the direction of electron motion) the electron has another definite character (or state function); this is its magnetic moment (or spin), which can be oriented in either of two directions in a magnetic field. Each direction is a different *state* and so a spatial state designated by a particular value of k can have each of the two spin states associated with it. The total number of states is therefore *double* the number of k states. This is similar to the situation which arose for the number of lattice vibrational states. This was trebled in order to take account of the three types of phonon polarization (section 5.10, p. 87).

Using (5.15) (but multiplying by two for the spin) the total number of electron states $G(k)$ from $k = 0$ up to a particular k is

$$G(k) = Vk^3/3\pi^2,$$

where V is the volume of the sample. In terms of the energy E, using (7.4) this is

$$G(E) = V(2mE)^{\frac{3}{2}}/3\pi^2\hbar^3. \tag{7.5}$$

The density of states $g(E)$ per unit energy interval is obtained by differentiating (7.5),

$$g(E) = V(2m)^{\frac{3}{2}}E^{\frac{1}{2}}/2\pi^2\hbar^3 = \tfrac{3}{2}G(E)/E. \tag{7.6}$$

It should be noted that the energy of an electron is proportional to k^2 (7.4), whereas for phonons it is proportional to k.† Eqns (7.5) and (7.6) therefore are not the same as the corresponding expressions for phonons.

We see from (7.5) that the greater the volume the more states can be accommodated up to an energy E. This is a direct consequence of the Heisenberg uncertainty relation—the greater the uncertainty in position, the less uncertainty will there be in the momentum (and the kinetic energy).

---

† For a phonon the energy $= \hbar\omega = \hbar k \times$ velocity if it is assumed that there is no dispersion.

7.4. The maximum energy of the electrons

We have now derived the density of electron states and we assume that the free electrons in the crystal are associated with these states. At 0 K the first electron will be in the lowest energy state, and subsequent electrons will occupy the next higher states, so that if there are N electrons altogether the N lowest energy states will be filled up to an energy E_{max}. Using (7.5) we may write

$$N = V(2mE_{max})^{\frac{3}{2}}/3\pi^2\hbar^3 \tag{7.7}$$

and

$$g(E_{max}) = \tfrac{3}{2}N/E_{max}. \tag{7.8}$$

We need to have some idea of the average energy of these electrons, but for simplicity let us first calculate the maximum energy. For a monovalent metal, such as sodium or copper, the electrons in the atom are arranged in closed shells with *one* extra electron on the outside. It is this one electron per atom which is the 'free' electron. In a mole of material containing 6×10^{23} atoms, this will also be N, the number of free electrons. If the molar volume is $\sim 10\,cm^3$ then we can substitute in (7.7) and calculate the maximum energy E_{max} which is necessary so that all N electrons may occupy permitted states. It comes out to be 4–5 eV or about 10^{-18} J. This should be compared with the energy of a classical particle which is of the order of kT. At room temperature it is worth remembering that kT is equivalent to $\frac{1}{40}$ eV (4×10^{-21} J). Thus E_{max} for the electrons in a metal is far higher than any classical energy. It would in fact correspond to the classical energy of a system at about 50 000 K! No wonder then that the classical theory does not work for electrons. If E_{max} is equated to the kinetic energy $\tfrac{1}{2}mv^2$ we find that the velocity of the electrons is about $10^6\,m\,s^{-1}$.

7.5. Distribution of the electrons amongst the energy states

At 0 K the N states of lowest energy will be populated as discussed in the previous section. How does the distribution change as the temperature is raised? Any variation in the electron energy must be provided by the thermal energy of the crystal. This, as we have just seen, is so small compared with the electron energy that the energy distribution can only change very slightly at any easily attainable temperature. From (7.6) we see that the distribution as a function of energy will have a parabolic form. At 0 K this will have a sharp cut-off at E_{max} (~ 5 eV) (Fig. 7.2(a)). As the temperature is raised, only electrons with energies near E_{max} can change their state because the thermal energy will be insufficient to excite electrons with lower energies to vacant states above E_{max}. The energy distribution therefore becomes slightly rounded at the cut-off and it has a thin tail extending to higher energies (Fig. 7.2(b)). This effect becomes more pronounced as the temperature is raised further.

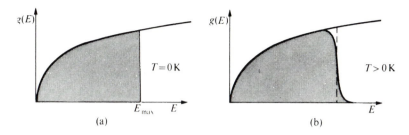

FIG. 7.2. The density of states $g(E)$ as a function of energy for the free-electron model. (a) At 0 K all states are occupied up to E_{max}. (b) At higher temperatures the occupation of states in the region of E_{max} is smeared out.

7.6. Fermi–Dirac statistics

We shall now put the results of this discussion into a more formal framework in order to obtain an expression for the number of electrons which have energies lying within a certain range when the material is at a temperature T.

In section 5.2 (p. 79) we introduced the Fermi–Dirac function $f_{FD}(E)$ which gives the probability that an electron (or any other indistinguishable particle of half-integral spin) has an energy E at a temperature T. This has the form

$$f_{FD}(E) = \left[\exp\left\{ \frac{(E - E_F)}{kT} \right\} + 1 \right]^{-1}. \tag{7.9}$$

As in section 5.3 the number of states between E and $E + dE$ which are occupied by electrons is

$$N(E)\, dE = g(E) f_{FD}(E)\, dE. \tag{7.10}$$

Now inspection of (7.9) shows that at 0 K $f_{FD}(E)$ is unity up to $E = E_F$ but that it is zero thereafter. Hence (7.10) will be zero for $E > E_F$. Since, as we have shown in section 7.3, $g(E)$ is a continuous function of E, the highest occupied electron state will have an energy E_F. Thus, at 0 K, $E_F = E_{max}$ (7.7). The form of $f_{FD}(E)$ is shown in Fig. 7.3(a). At 0 K it falls discontinuously to zero when $E = E_F$, but for $T > 0$ K the drop at E_F is smeared out. The calculation of $N(E)$ is illustrated (Fig. 7.3(a), (b), (c)).

It should be noted particularly that $f_{FD}(E)$ gives the probability of occupation of a state with energy E but it gives no indication as to whether such a state with that energy actually exists. This is taken care of by the term $g(E)$. In the free electron theory $g(E)$ is a continuous function and no problems arise, but for semiconductors in particular there are energy gaps in which $g(E)$ is zero but in these regions $f_{FD}(E)$ is still continuous and has non-zero values. Note that when $E = E_F$, $f_{FD}(E) = \frac{1}{2}$.

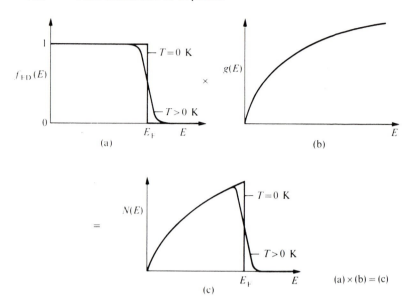

FIG. 7.3. The calculation of the density of occupied electron states $N(E)$. (a) The Fermi–Dirac function; (b) the density of states; (c) $N(E) = g(E) \cdot f_{FD}(E)$. The shaded regions of Fig. 7.2 correspond to (c).

E_F is called the Fermi energy. Its value is formally defined by equating the total number of occupied states to the actual number of free electrons. This normalizes the function $f_{FD}(E)$, i.e.

$$N = \int_0^\infty g(E) f_{FD}(E)\, dE.$$

In metals, however, it is usually sufficient to assume that E_F is equal to E_{max} and to ignore any temperature-dependence. In semiconductors it should be noted that E_F is *not* in general equal to E_{max}.

7.7. The mean energy of the electrons

The total energy E_{tot} of all the free electrons can be calculated by multiplying the population of each state by its energy and then summing (or integrating) over all states, i.e.

$$E_{tot} = \int_0^\infty E g(E) f_{FD}(E)\, dE. \tag{7.11}$$

The average energy per electron $\langle E \rangle$ is then E_{tot}/N,

$$\langle E \rangle = 1/N \int_0^\infty Eg(E)f_{FD}(E)\,dE. \qquad (7.12)$$

Integrals involving $f_{FD}(E)$ tend to be unpleasant, and treatments are given in more advanced texts. We shall avoid a detailed mathematical analysis by using simple approximations. In the present case we shall evaluate $\langle E \rangle$ at 0 K; $f_{FD}(E)$ is then unity up to E_F, beyond which it is zero. From (7.6) we can write $g(E)$ in the form $AE^{\frac{1}{2}}$, and using (7.12) we then have

$$\langle E \rangle = (A/N) \int_0^{E_F} E^{\frac{3}{2}}\,dE$$
$$= (\tfrac{2}{5})(A/N)E_F^{\frac{5}{2}}. \qquad (7.13)$$

But from (7.6), at 0 K we have

$$N = \int_0^{E_F} g(E)\,dE = A \int_0^{E_F} E^{\frac{1}{2}}\,dE,$$

i.e.

$$N = (\tfrac{2}{3})AE_F^{\frac{3}{2}}, \qquad (7.14)$$

and so substituting this for N in (7.13) we obtain

$$\langle E \rangle = \tfrac{3}{5}E_F. \qquad (7.15)$$

A more detailed calculation shows that $\langle E \rangle$ changes very little with temperature. The mean energy of the electrons is therefore always close to $\frac{3}{5}E_F$, i.e. a few electron volts.

7.8. The electronic specific heat

We have already discussed why, as the temperature is increased, only those electrons which have energies which are very close to E_F are able to change their state. There is therefore only a very slight increase in the electron-energy distribution (Fig. 7.2(b)), and this means that the electronic specific heat is very small. Thus, as mentioned in the introduction to this chapter, the contribution of the electrons to the total specific heat of a metal is undetectable at room temperature.

Formally, the electronic specific heat c_e can be written as $\partial E_{tot}/\partial T$, where E_{tot} is defined by (7.11). A simple solution,† however, may be obtained by assuming that only those electrons which have energies lying within the thermal energy ($\sim kT$) of E_F can change their states as the temperature is increased above 0 K (Fig. 7.4). The actual number of electrons involved will be the density of states per unit energy range at E_F, multiplied by the energy range kT itself, i.e. $g(E_F)kT$. Each of these electrons is able to interact with the

† See Problem 7.8 for a more detailed treatment.

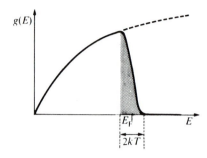

FIG. 7.4. Calculation of the electronic specific heat. Only electrons with energies in the neighbourhood of E_F can change their state as the temperature is raised, and so the electronic specific heat is small.

phonons and thereby increase its energy by $\sim kT$. The total increase in the energy of the electron system above its energy at 0 K is therefore equal to the number of electrons involved $\times$ the energy each one can gain, i.e.

$$g(E_F)k^2T^2.$$

The electronic specific heat is obtained by differentiating this extra energy with respect to T, i.e.

$$c_e \approx 2g(E_F)k^2T. \tag{7.16}$$

The more accurate expression obtained by differentiating (7.11) is

$$c_e = \tfrac{1}{3}\pi^2 g(E_F)k^2T \equiv \gamma T. \tag{7.17}$$

It is sometimes useful to use (7.8) and write

$$c_e = \tfrac{1}{2}\pi^2 \frac{N}{E_F}k^2T. \tag{7.18}$$

The electronic specific heat is usually written in the form $c_e = \gamma T$, where γ is proportional to the density of states at the Fermi energy. If we use (7.18) with $E_F = 5$ eV then at 300 K c_e is about 0.2 J K^{-1} mol^{-1}, which is only about 1 per cent of the lattice specific heat (25 J K^{-1} mol^{-1}). But at low temperatures c_e becomes quite important because then the lattice specific heat c_g has the form AT^3 (eqn (5.25)), and so this decreases much more rapidly than does c_e. In the liquid-helium region the two specific heats are usually comparable with one another (Fig. 7.5(a)): we may write for the total specific heat,

$$c = \gamma T + AT^3. \tag{7.19}$$

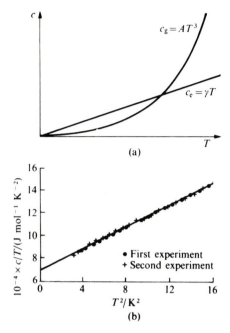

(a)

(b)

FIG. 7.5. The specific heat of metals at low temperatures. (a) In the helium region the electronic contribution c_e can dominate that of the lattice, c_g. (b) A plot of c/T against T^2 for copper, showing the very good linear relationship which is obtained (C. A. Bailey (1959), D. Phil. Thesis, Oxford University).

and hence a plot of c/T against T^2 should be a straight line with an intercept of γ. An example is shown in Fig. 7.5(b). From such an experimental determination of γ we may then deduce from (7.17) a value for $g(E_F)$ which can be compared with that calculated from (7.7) and (7.8). The agreement is never entirely satisfactory but it does give an indication as to whether the free-electron theory is a good model for the metal in question.

The success in explaining why the electronic specific heat is so small is undoubtedly the major triumph of the free-electron theory. There are, however, some other phenomena which can also be treated with this model and they are discussed briefly in the following sections.

7.9. Free-electron paramagnetism

When materials are placed in a magnetic field they acquire an induced magnetic moment. This may be either in such a direction as to oppose the applied field (diamagnetism) or it might be parallel to the field (paramagnetism and ferromagnetism). Diamagnetism is present in all materials. In the absence of any other masking effect it is always observed. Strong paramagnetism, which is temperature-dependent, and ferromagnetism occur in some elements and compounds containing certain transition-group ions of the periodic table; these are discussed in Chapters 11 and 12. All metals, however, show a weak paramagnetic effect which is not temperature dependent and this may be explained using the free-electron theory which we have developed.

This paramagnetism (sometimes called Pauli paramagnetism) arises from the magnetic moment associated with the intrinsic spin of the free electrons. In zero magnetic field these moments have no special orientation, but in an external field B quantum mechanics predicts that they may only have either of two orientations—they may be parallel or antiparallel to B. Since the energy of a dipole which is parallel to B is less than that for one which is antiparallel, there will be more parallel dipoles than antiparallel ones because this will reduce the energy of the system. There will therefore be a net magnetic moment parallel to B and this is the paramagnetic effect.

If the magnetic moment of an electron is m_m its magnetic energy will be $-m_m B$ when it is parallel to B and $+m_m B$ when it is antiparallel to B. The energy distribution and occupation in zero field, shown in Fig. 7.6(a) will now be shifted to higher and lower energies by $m_m B$ for the antiparallel and parallel alignments respectively. These are shown as the upper and lower parts of Fig. 7.6(b). It is clear from this diagram that electrons in the top half of the figure (antiparallel) can reduce their energy by reversing their orientation, thereby occupying vacant states at the top of the parallel distribution. The number which can do this advantageously will be those which are within $m_m B$ of the original Fermi energy of the top part, i.e. $\frac{1}{2}g(E_F)m_m B$ (the factor $\frac{1}{2}$ is needed because we have divided the energy distribution into two parts). The figure shows a sharp cut-off in occupation at E_F because for simplicity we have assumed $T = 0$ K.

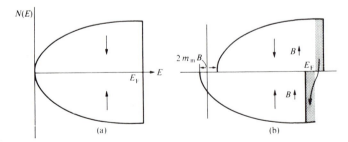

FIG. 7.6. Free-electron paramagnetism. (a) In zero magnetic field there are equal numbers of electrons with opposing spins. (b) In a field B the electron distributions for the two spins are shifted relative to one another and the energy of the system can be reduced by a reversal of some spins near E_F.

The difference in population between the two groups of electrons will therefore be $g(E_F)m_mB$, and so the net magnetic moment is $g(E_F)m_m^2B$. The paramagnetic susceptibility χ_p is defined as magnetic moment/H, where H is the magnetic intensity ($= B/\mu_0$).
Hence

$$\chi_p = g(E_F)m_m^2\mu_0. \tag{7.20}$$

We therefore see that, as in the case of the electronic specific heat, χ_p for the free electrons is almost entirely due to the relatively small number of electrons with energies close to E_F which are able to change their energies. Above 0 K we should take account of the tail of the Fermi–Dirac function around E_F. This does give a very slight temperature dependence but for all practical purposes χ_p does not vary with T. It is very small, being of the order of 10^{-5} MKS m^{-3}.†

It should be noted that the diamagnetism, mentioned at the beginning of this section, tends to partially cancel χ_p to about two-thirds of the value predicted by (7.20).

Electron emission

7.10. Thermionic emission; the work function

Whilst the vacuum tube has not retained its pre-eminent position in electronics, the emission of electrons from a hot filament is still a very important phenomenon which has many applications. Since electrons are retained within metals and do not spill out very easily, their energies within the metal must be smaller than when they are outside. There must therefore be a potential barrier at the metal surface which in terms of the free-electron theory can be thought of as being dependent on the height of the sides of the box which contains the electrons.

The *work function* ϕ is the minimum extra energy‡ (measured above E_F) which an electron must have in order for it to be able to escape from the metal to infinity. It can be considered as being due to the fact that, when an electron outside the metal is at a distance x from the surface, it experiences an image force $-e^2/(16\pi\varepsilon_0 x^2)$ which attracts it back to the surface. In order that it can escape completely it must have enough energy to overcome this force and this energy is the work function.§ The potential energy of the electron will therefore vary with x as is shown in Fig. 7.7. Since the concept of an image force

---

† *Classically* the population of parallel to antiparallel states should be as $1 : \exp(-2m_mB/kT)$. This gives a much larger value for χ_p which is strongly temperature dependent.

‡ It should be noted that in some texts ϕ is quoted in volts. It is then a potential, not an energy.

§ This is the basic principle, but the surface condition and the presence of positive and negative charge layers at the surface strongly influence the value of the work function.

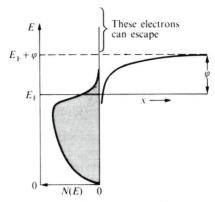

FIG. 7.7. The work function. Only those electrons with an energy which is greater than $E_F + \phi$ can escape from the surface of the metal.

is only justified if the medium is continuous the simple expression for the force is not applicable when the electron is very close to the surface and so it cannot be used as a basis for a quantitative estimate of the work function.

The minimum energy which an electron must have in order that it can be emitted from the metal is $E_F + \phi$. Such electrons will be in the high-energy tail of the electron distribution and so as the temperature is raised the emission will be increased. If an electrode which is at a sufficiently high positive potential is placed near the metal it will attract all the electrons which are emitted and we may then measure the saturation current at that temperature.

In the next section we shall see that the work function and hence the saturation current depends slightly on the potential, but we shall ignore this for the moment.

In order to calculate the saturation thermionic current we need to find the number of electrons which have energies greater than $E_F + \phi$. There is, however, a slight complication because we must take account of the fact that only those electrons which have a sufficiently large component of their momentum (say p_x) *normal* to the surface which we are considering can escape from that surface, i.e. we need $p_x^2/2m \geqslant E_F + \phi$. The other components of the momentum p_y and p_z may have any value. The complete calculation is simple but tedious and may be found in many texts. The expression for the saturation current density j which results is

$$j = emk^2T^2/(2\pi^2\hbar^3) \exp(-\phi/kT) = AT^2 \exp(-\phi/kT). \qquad (7.21)$$

This is the Richardson Dushman equation. A is about $1 \cdot 2 \times 10^6$ A m^{-2} K^{-2} by calculation and between $0 \cdot 7 \times 10^6$ A m^{-2} K^{-2} and 1×10^6 A m^{-2} K^{-2} by experiment. The temperature dependence is dominated by the exponential term and so the T^2 part cannot be verified directly. ϕ has values of about 4 eV for most metals, but for various oxide-coated filaments it is about 2 eV.

7.11. Field emission; the Schottky effect

We must now see how an external electric field or potential affects the work function.

When an electrode which is used to collect the electrons is charged positively to a potential V (this is, of course, a negative potential for electrons) there is an electric field between the metal and the electrode in such a direction that the electron is attracted towards the electrode, i.e. the potential energy of the electron decreases towards the electrode. This is shown by curve A in Fig. 7.8. To this potential must be added the original image-charge potential which was discussed in the previous section and which is redrawn as curve B.

The total potential energy is the sum of contributions A and B. As can be seen this has a maximum value which will be lower than the original ϕ by an amount $\Delta\phi$. It is clear that the higher the field (i.e. V) the greater is $\Delta\phi$. This effect is a small one, e.g. if the field is $\sim 10^4\,\mathrm{V\,m^{-1}}$, $\Delta\phi$ is about $10^{-2}\,\mathrm{eV}$.

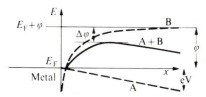

FIG. 7.8. The Schottky effect. If there is a suitable potential gradient (curve A) outside the metal the energy of the electron is given by curve A + B and the work function is reduced by $\Delta\phi$.

7.12. High-field emission

If, however, the potential V in the preceding discussion is very high, an interesting new effect comes into play. The potential energy then drops so sharply that its value at a short distance Δx from the surface has fallen by an amount of the order of ϕ (Fig. 7.9), so that the energy of an electron at Δx is equal to that of an electron within the metal which has an energy E_F. If Δx is sufficiently small ($\sim 10^{-9}$ m) the electrons can then 'tunnel'† through the potential barrier to a position outside which has the same energy. They do not have to have an energy $E_F + \phi$ to escape from the metal.

† Tunnelling is dealt with in all quantum mechanics texts. It may be understood as being due to the fact that the electron waves in the box do not cut off sharply at the box boundaries (if the boundaries are not infinitely high) but they fall to zero within a short distance outside them. This means that within this distance there is a small but finite probability that an electron will be found outside the box, provided that the mean energy, inside and outside the box, is conserved. This is called tunnelling.

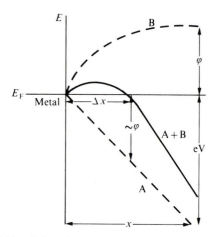

FIG. 7.9. High-field emission. If the potential V is sufficiently large the curve A + B (see also Fig. 7.8) is sufficiently depressed that it has a value equal to E_F at a distance Δx from the metal surface which is so small that electron tunnelling can occur across Δx.

If in Fig. 7.9 ϕ is 4 eV, Δx is 10^{-9} m, and a potential V is applied between the metal and an electrode 1 cm away from the surface, then a simple calculation using similar triangles shows that for the energy at Δx to be equal to E_F (i.e. for tunnelling to occur) requires V to be $4 \times 10^{-2}/10^{-9} = 4 \times 10^7$ V. Surface irregularities are usually greater than 10^{-9} m, and in practice emission may be observed if V is 10^4 to 10^5 V, i.e. a field of 10^6 to 10^7 V m^{-1}.

This is the mechanism which is responsible for spark discharges from sharply pointed electrodes.

7.13. The field-ion microscope

An interesting development of high-field emission techniques is the field-ion microscope.† A very fine metal needle (e.g. of tungsten) with a hemispherical tip is cooled with liquid helium (or nitrogen) and is maintained at a high positive potential with respect to a thin conducting layer and fluorescent screen (Fig. 7.10(a)). Very low-pressure helium gas is admitted to the vacuum space. The atomic arrangement at the tip of the needle makes certain atoms more prominent than others (Fig. 7.10(b)), and the field gradient near them is

---

† This is not a direct application of electron emission although the original field-emission microscope from which the field-ion microscope was developed did use electron emission from the most prominent atoms to show their positions on a screen. The resolution was much poorer because (1) the de Broglie wavelength of electrons is longer than that of He ions and (2) the needle was heated by the electron current which flowed in it.

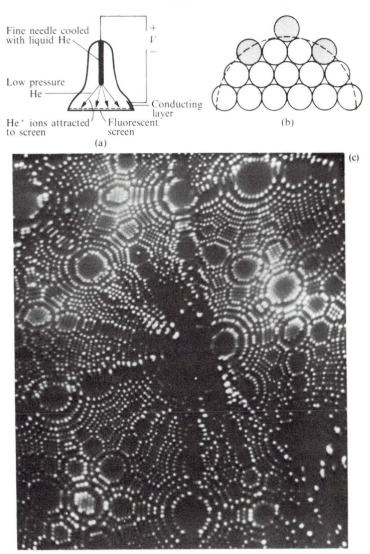

Fine needle cooled with liquid He

+
V
−

Low pressure He

Conducting layer

He⁺ ions attracted to screen

Fluorescent screen

(a)

(b)

(c)

FIG. 7.10. The field ion microscope. (a) Diagram showing the main features. (b) The atoms at the hemispherical tip of a fine needle showing how some are more prominent than others. (c) A field ion microscope picture of tungsten. A grain boundary runs from top centre to slightly right of bottom centre as can be seen from the change in the atomic arrangement. The diameter of the region covered in the picture is about 100 nm. (Photograph by courtesy of T. G. Godfrey, D. A. Smith, and G. D. W. Smith.)

increased. The effective field can be greater than 10^{10} V m^{-1}. The helium atoms are ionized at these atoms, they become positively charged, and they travel to the screen. In this way the positions of the most prominent atoms in the needle tip are displayed (Fig. 7.10(c)).

7.14. Contact potential

If two dissimilar conductors are connected together their Fermi energies will not in general be coincident. Electrons therefore will flow from one material to the other so that the energy of the system is minimized (in an analogous way to that suggested for Pauli paramagnetism in section 7.9, p. 116). This tends to equalize the Fermi energies, and the electron flow charges up one conductor relative to the other so that there will be a *contact potential* between them (Fig. 7.11). It is important to note that when equilibrium has been established the Fermi energy throughout the system is the same. (See section 10.1, p. 160.) This equalization of the Fermi energies is particularly important in the theory of semiconducting junctions.

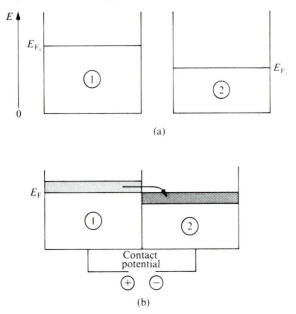

Fig. 7.11. Contact potential. (a) Two dissimilar metals will in general have their occupied electron states filled up to different energies and their Fermi energies will therefore not be the same. (b) If the metals are placed in contact the energy of the combined system can be reduced by a transfer of electrons from one metal to the other. The charge that is built up gives rise to a contact potential between them. N.B. The fraction of the electrons which need to be transferred is greatly exaggerated in this diagram, see section 10.2.

PROBLEMS

7.1. Calculate the Fermi energy. E_F. at 0 K for potassium (atomic weight = 39. density = 860 kg m^{-3}). Determine the density of states at E_F for (a) a specimen of volume 1 cm^3. and (b) a g mole. Explain why the values from (a) and (b) are not the same. What proportion of the free electrons have energies within kT of E_F at 300 K?

7.2. The Fermi energy of copper at 0 K is 7 eV. Calculate the mean energy of the conduction electrons and their root mean square velocity.

7.3. Calculate the electronic specific heat of a g mole of copper at 300 K. At what low temperature are the electronic and lattice specific heats of copper equal to one another? (Debye θ of copper = 348 K.)

7.4. When a metal is compressed the Fermi enery is increased. Explain why this is so. Show that the compressibility, $-1/V(dV/dP)$, of the conduction electron system is given by $3V/(2NE_F)$, where N is the number of electrons in a volume V. Calculate the value of the compressibility of copper and compare your answer with the experimental value of 7×10^{-12} N^{-1} m^2. Explain why the two values do not agree. (For Cu, atomic weight = 63·5, density = 8900 kg m^{-3}.)

7.5. Estimate the paramagnetic contribution to the magnetic susceptibility per m^3 of potassium, for which the Fermi energy is 2·1 eV.

7.6. A metal contains 5×10^{28} free electrons per m^3. Estimate the contact potential which would exist between two samples of this metal if one were subjected to a pressure of 30 atmospheres and the other was maintained at atmospheric pressure. assuming that the bulk modulus is 10^{11} N m^{-2}. (1 atmosphere = 10^5 N m^{-2}.)

7.7. A free-electron system in which each level is 6-fold degenerate has its energy levels filled up to $k = (2\pi^2)^{\frac{1}{3}}/a$. where a is the interatomic spacing. Calculate the number of conduction electrons per atom.

7.8 Derive an expression for the electronic specific heat as follows. Substitute for $g(E)$ and $f_{FD}(E)$ [using eqns (7.6) and (7.9) respectively] into eqn (7.11) and differentiate with respect to T. Write a programme to calculate the integral and compute the specific heat (at 100, 200 and 300 K) for a monovalent metal which has a kg molar volume of 10^{-2} m^3. You will need to calculate E_F. In order to save computer time consider carefully the limits of integration that you should use.

8. Electrical conductivity and the band theory

8.1. Elementary conductivity theory

IN this chapter we describe the extension which must be made to the free-electron theory which enables us to understand the behaviour of the electrical conductivity of both metallic and non-metallic crystals.

If a conductor is in zero electric field the electrons are distributed amongst all the energy states, up to E_{max} (or E_F) as discussed in the previous chapter. There will therefore be as many states occupied which have a positive wave vector as there will be those with a negative wave vector, i.e. there will be as many electrons travelling in one direction as in another. If we sum over all the electrons, their average velocity is zero and so there is no net electric current.

This is illustrated in Fig. 8.1(a), which shows an isotropic distribution of wave vectors (corresponding to occupied states) out to some maximum values k_F, which lie on the surface of a sphere (called the Fermi surface). This is a constant-energy surface in k-space, corresponding to the Fermi energy E_F.

When an electric field $\mathscr{E}$ is applied to the conductor each electron will experience a force[†] $-e\mathscr{E}$ and it will undergo an acceleration in the direction of that force (for electrons this direction is of course in a direction opposite to that of the conventional field). The effect of the field is shown in Fig. 8.1(b). The distribution of occupied states has now become unbalanced, and there are now more electrons travelling to the right than to the left. This implies that there will be a net current flow. However, the story is not yet complete because if the field was maintained the electrons would continue to accelerate and so the sphere in Fig. 8.1(b) would become more and more off-centred, i.e. the current would continue to increase indefinitely. But we know from experience that in a very short time a constant current is established whose value depends on the field. This stabilization occurs because electrons in states near the Fermi surface are scattered by various mechanisms (which we shall discuss later) to vacant states of lower k-value and energy. The most accessible vacant states will be on the left hand side of the Fermi surface in Fig. 8.1(b). A dynamic equilibrium is therefore established between the electrons which are being continually accelerated to the right-hand and those which are scattered back to the left-hand side of the figure.

[†] We shall use e for the modulus of the electronic charge, so that the charge on the electron is $-e$.

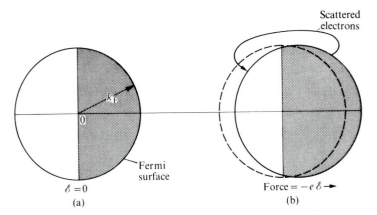

FIG. 8.1. Occupied electron states in k-space for a simple free-electron system. (a) If there is no external electric field all states are filled within a sphere of radius k_F, centred on $k = 0$. (b) In an electric field the spherical distribution is shifted in the direction of the force on the electrons and so there are more of them travelling to the right than to the left. There is a net electron current. Note that in practice the Fermi surface is not usually spherical—see sections 8.11, 8.12.

8.2. The electron mean free path

We assume that the electrons travel a distance l (the mean free path) between each scattering. The velocity v_F at the Fermi surface is very high, $\sim 10^6$ m s^{-1} (section 7.4, p. 110), compared with any change in velocity produced by the applied field, and so the time between collisions is more or less constant at l/v_F. This means that the amount of offset in Fig. 8.1(b) will be greater as the field is increased because a larger change of velocity can be produced in the same time l/v_F. The current will therefore be higher and this, of course, is the basis of Ohm's law.

The electrical conductivity σ is often defined in terms of the current density j and the field $\mathscr{E}$ by

$$j = \sigma\mathscr{E} = \mathscr{E}/\rho, \qquad (8.1)$$

where $\rho = 1/\sigma$ is the resistivity.

A kinetic theory argument† gives the following expression for the resistivity,

$$\rho = mv_F/Ne^2l, \qquad (8.2)$$

† A simple derivation is as follows: equate the force $e\mathscr{E}$ on an electron to the mass × acceleration, $|e\mathscr{E}| = m\,dv/dt$. Integrate over the period τ between two collisions, $e\mathscr{E}\tau = m(v_1 - v_2) = mv_d$, where v_d is the extra drift velocity. But $\tau = l/v_F$ and therefore $v_d = e\mathscr{E}l/mv_F$. Since $j = Nev_d$ we can substitute for v_d, and so $j = Ne^2\mathscr{E}l/mv_F$. Then by comparison with (8.1) $\rho = mv_F/Ne^2l$.

where N is the number of conduction electrons per unit volume. The resistivity of copper at room temperature is $\sim 2 \times 10^{-8}\,\Omega$ m, v_F is $\sim 10^6$ m s^{-1}, and N is $\sim 7 \times 10^{28}$ m^{-3}. If these values are substituted in (8.2) we find that l is $\sim 2 \cdot 5 \times 10^{-8}$ m, i.e. the electrons travel about 100 atomic (or ionic) spacings between collisions. Since one might think that each positive ion would play a part in scattering the electrons, this rather long mean free path is surprising; but in fact, as we shall explain, ionic scattering is *not* an important factor in determining the resistivity.

The apparent inability of the ions to scatter electrons cannot be accounted for on the basis of the free-electron theory. This is hardly surprising, since there we assumed that the electrons move in an empty box. The model must be developed further and it is necessary to take account of the fact that the electrons are moving through a region containing a *regularly arranged* lattice of positive ions. These ions produce a pattern of periodically varying electric fields (or potentials), and as the electrons travel within the metal they experience the influence of this periodic potential $V(r)$.

The problem is analogous to that of X-ray diffraction in a crystal except that now we are interested in the effect of *electron* waves which interact with a periodic lattice. The basic result is very similar to the X-ray case. The waves are *not* scattered by the lattice if the lattice is *perfectly* periodic, except for some very special values of the wave vector (when k satisfies the Bragg condition). This may be explained in a simple manner by assuming that if the lattice is regular an electron is attracted to the ions as much in one direction as it will be in the opposite direction. If the lattice contains imperfections the Bragg condition still holds but there will be in addition a general diffuse background scattering of radiation. In the electron case it is this which gives rise to a finite mean free path. In order to discuss the electrical resistance we must therefore study the ways in which the periodicity of the lattice may be upset.

8.3. Electron-scattering mechanisms

There are two main types of lattice imperfection which can contribute to the electrical resistivity:

(1) the thermal vibration of the lattice will prevent the atoms from ever all being on their correct sites at the same time;

(2) the presence of impurity atoms and other point defects will upset the lattice periodicity.

The lattice vibrations decrease as the temperature is reduced and we would therefore expect that their contribution to the resistivity, usually denoted by ρ_i,† will decrease at low temperatures, eventually becoming zero at 0 K.

---

† The subscript to ρ_i stands for 'ideal', as this resistivity must be present even in an ideally pure crystal.

At low temperatures ρ_i varies as a high power of T which is often quoted as T^5, but this precise temperature dependence is not usually observed. At higher temperatures ρ_i becomes linear with T. This is shown in Fig. 8.2.

The larger the amplitude of vibration at any temperature, the greater will be ρ_i. Since this amplitude depends on the interatomic forces and hence on the inverse of the Debye θ (section 5.13, p. 89) it is to be expected that ρ_i will be less for metals with a high θ, and vice versa, and this is confirmed by experiment.

The arrangement of point defects in a crystal does not change with temperature and therefore the amount of scattering and resistivity ρ_0 which they produce would be expected to be constant. Their contribution to the resistivity is temperature independent but it does of course increase with the impurity concentration.

The total resistivity ρ is therefore

$$\rho = \rho_0 + \rho_i. \tag{8.3}$$

This is also drawn in Fig. 8.2, in which it can be seen that ρ at first decreases linearly with T and at low temperatures it flattens off to a constant value, equal to ρ_0, which is called the *residual resistance*. It is clear that for a very pure sample ρ_0 will be very small, whereas for an impure specimen it will have a high value.

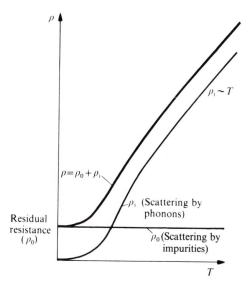

FIG. 8.2. The electrical resistivity, ρ, is the sum of two contributions. A constant ρ_0 due to scattering by impurities and ρ_i which is caused by electrons being scattered by phonons.

8.4. The residual resistance ratio

At room temperature ρ_i dominates and so the resistivity, ρ_{RT}, is almost the same for all samples. This means that the ratio ρ_{RT}/ρ_0, which is called the residual resistance ratio, can be used as a measure of the purity of a sample. ρ_0 can be measured by immersing the specimen in a bath of liquid helium, because the resistivity of nearly all metals is in the residual range by that temperature. The values of ρ_{RT}/ρ_0 which can be obtained will of course depend on the particular metal which is being studied, but for high-purity samples of metals such as aluminium and copper the ratio can be ~ 200, although for very carefully prepared materials it is possible for the ratio to be as high as 10^4–10^5. The measurement of this resistance ratio is a very straightforward method for determining the quality of a particular sample although of course it gives no indication of the *nature* of the impurities or the defects.

The simple addition of ρ_0 and ρ_i in (8.3) is often referred to as Mattheissen's rule. It is valid if ρ_0 and ρ_i are of different magnitudes, but it may be slightly in error (by a few per cent) if they have similar values.

8.5. Transport of heat in metals

In metals the heat is carried almost entirely by the electrons because the phonons are so strongly scattered by the many electrons present that the direct transportation of thermal energy by the phonons, which we discussed in Chapter 6 is very small indeed. The thermal resistance is therefore also controlled by the electron-scattering mechanisms which we have just discussed and we would expect there to be a close relationship between the electrical and the thermal conductivities. This is provided by the Wiedemann–Franz law.

8.6. The Wiedemann–Franz law

If we insert into the kinetic theory formula for κ (6.1) the expression for the electronic specific heat (7.17), we obtain

$$\kappa = \tfrac{1}{3}\gamma T v l'. \tag{8.4}$$

We may write γ as $\pi^2 N k^2 / 2 E_F$ from (7.18), and so

$$\kappa = \pi^2 N k^2 v l' T / 6 E_F.$$

Using (8.2) for the electrical conductivity σ we may calculate the ratio

$$\frac{\kappa}{\sigma T} = \frac{\pi^2 N k^2 v l' m v_F}{6 N E_F e^2 l}.$$

If $v \approx v_F$, so that $m v v_F \approx 2 E_F$, and if $l = l'$ then this expression reduces to

$$\frac{\kappa}{\sigma T} = \frac{\pi^2 k^2}{3 e^2} = 2 \cdot 45 \times 10^{-8} \ \text{W} \, \Omega \, \text{K}^{-2} \equiv L, \text{ the Lorenz constant.} \tag{8.5}$$

This relationship between the electrical and thermal conductivities of a metal holds remarkably well. It is only in very pure specimens at intermediate temperatures, where small angle phonon scattering is important, that it is not very satisfactory. This is because at each electron/phonon interaction charge is conserved, whereas thermal energy (of the electron) is not. This has the effect of making l larger than l'.

8.7. The thermal conductivity

Although the heat and the electrical currents are both transported by the electrons, which usually have the same value of l for both mechanisms, the actual temperature-dependence of the thermal resistivity W is *not* the same as that for the electrical resistivity. This is because in an electric current the charge transported by an electron is constant (equal to the electronic charge), whereas for heat conductivity the thermal energy which can be transported depends on the thermal energy an electron is able to transfer at any given temperature. This depends on its specific heat which, as can be seen from (8.4), itself varies with temperature (γT). This is why the Wiedemann–Franz law (8.5) contains a T in the denominator and it is *not* just $\kappa/\sigma = $ constant.

From our knowledge of the electrical resistivity we can deduce the behaviour of the thermal resistivity. At high temperatures, where lattice vibrational scattering is dominant, we have seen that ρ (i.e. mainly ρ_i) varies as T and at low temperatures in the impurity scattering region it is constant (i.e. mainly ρ_0). From (8.5) we can see therefore that the thermal resistance at high temperatures due to vibrational scattering W_i will be constant (and will fall to zero at low temperatures) whilst the impurity scattering term W_0, which is dominant at low temperatures, will be proportional to T^{-1}. These curves are shown in Fig. 8.3(a). In the intermediate region where, as already stated, (8.5) does not hold very well, W_i varies as T^2.

The total thermal resistance W which is also shown in the figure passes through a minimum in the intermediate region at low temperatures. The purer the sample the deeper is this minimum. In the thermal conductivity diagram (Fig. 8.3(b)) this minimum becomes a maximum which is sharper and higher for purer specimens. Below this maximum κ is proportional to T and this linear dependence is well substantiated by experiment.

8.8. The band theory

In section 8.2, p. 126, we discussed how the presence of a *regular* periodic potential due to the positive ions in the lattice did not, in itself, give rise to electron scattering. This potential $V(r)$, however, does change the distribution of electron energy states and a development of the theory enables us to obtain a profound understanding of the nature of electrical conductors.

In principle the Schrödinger equation (7.1) to which the extra potential term $V(r)$ has been added, must be solved. This equation is then

$$\frac{-\hbar^2}{2m} \frac{\mathrm{d}^2\psi}{\mathrm{d}r^2} + V(r)\psi = E\psi.$$

(8.6)

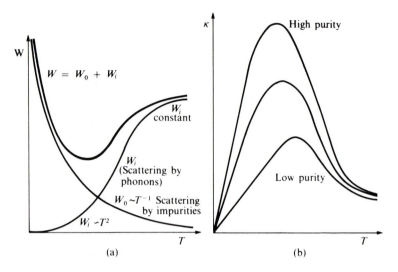

FIG. 8.3. Conduction of heat by electrons. (a) The electronic thermal resistivity W is the sum of two contributions, W_0 which increases as the temperature is reduced is caused by impurity scattering and W_i is due to electrons being scattered by phonons. (b) The thermal conductivity, κ, as a function of temperature for specimens of different purities.

$V(r)$ is a periodic function which has the same periodicity as the crystal lattice, i.e. $V(r) = V(na + r)$, where a is the lattice constant and n is an integer.

This equation, which was first treated by Bloch, is difficult to solve properly, and even approximate methods are troublesome. We therefore first give a description of the main features of the solution and discuss the very important consequences which stem from them. We postpone until section 8.15 *et seq.* an attempt to present some justification for the form of the solution.

The most important difference between the periodic-potential theory and the free-electron theory is that the relationship between E and $\boldsymbol{k}$ is no longer the simple parabolic function (7.4). Instead the function is multivalued and there are certain bands of energy which are forbidden, i.e. no permitted $\boldsymbol{k}$-states exist for them. The relation between E and $\boldsymbol{k}$ for one dimension of k is shown in Fig. 8.4(a).

The main feature which should be noted is that all the curves repeat themselves after an interval in k of $2\pi/a$ (a basic vector of the reciprocal lattice— section 2.7, p. 26). There is a fundamental similarity between this behaviour and that of phonons—we saw that a phonon with a long wave vector was equivalent in every physical way to one which differed from it by $2\pi/a$. In Fig. 8.4(a) the electron states on the bottom curve indicated by points A, A′, and A″ are identical states.

The upper curves show that for each wave vector there are several permitted energies—one within each energy band (e.g. B, B′, B″). In order to avoid the confusion that would arise if the complete multi-valued set of curves was used, it is conventional to use certain sections of them which are so chosen that as k increases so does the energy. The smallest values, starting from $k = 0$, are reserved for the lowest energy band (bottom curve); the next set from $k = \pi/a$ to $2\pi/a$ and $-\pi/a$ to $-2\pi/a$ is reserved for the second energy band (one curve higher), and so on. This is shown in Fig. 8.4(b). This is called the extended-zone scheme. It should be emphasized that this is merely a useful convention. Any sets of k which cover a range of $2\pi/a$ could be used to span all the available energy states. Another scheme (which saves space on diagrams) is the *reduced-zone scheme* shown in Fig. 8.4(c), where the same range of k from $-\pi/a$ to π/a is used to cover all energy bands.

It is important to note the comparison between the $E \sim k$ relationship for the free-electron model and for the extended-zone scheme. This is shown as a dashed curve in Fig. 8.4(b).

8.9. Metals, insulators and semiconductors

The number of permitted values of k in each zone is equal to N_i, the number of *ions*† in the crystal. As in the free-electron theory, each k-state may be associated with either of two electron spin states, and so the total number of states in each zone is $2N_i$. This means that if a metal has only one conduction electron per atom they will just occupy all the states in the lower half of the zone (Fig. 8.5(a)). We therefore see that for such a monovalent metal the energy distribution is virtually the same on the basis of either the free-electron or the band-theory models.

The important differences arise when there are an even number of electrons per atom because these would occupy an integral number of zones (Fig. 8.5(b)). If an electric field is now applied to drive a current through the crystal the shift in the electron distribution (Fig. 8.1(b)) cannot occur unless the zones are so close together that electrons at the top of the uppermost filled zone can obtain enough extra energy (e.g. from thermal excitation) to jump the forbidden energy gap so that they can occupy permitted energy states at the bottom of the next zone. Since this energy gap can be a few electron volts (i.e. high compared with thermal energies) excitation to the next band cannot occur and so no current can flow. The material would therefore be an *insulator*.

If the energy gap is not too large, say 1 eV or less, a few electrons can be excited across the energy gap at room temperature and so a small current can flow (Fig. 8.5(c)). The amount of excitation will increase as the temperature

† More rigorously, it is the number of primitive cells (section 1.7, p. 7).

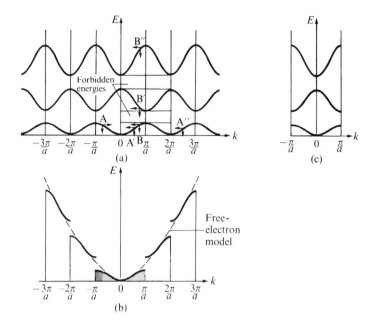

FIG. 8.4. The relationship between electron energy and wave vector for the simple band theory. (a) The extended-zone diagram which shows the full solution. (b) The more commonly used diagram which incorporates sections from the various curves of (a). (c) The reduced zone scheme in which the curves of (b) are folded back into the first zone.

is raised so that more electrons can be excited to the next band. The conductivity therefore *increases* with temperature. This is the behaviour which is typical of a *semiconductor*. It will be recalled that in a metal the conductivity *decreases* with temperature (section 8.3).

We therefore see that the band model provides a comprehensive explanation for the three types of electrical behaviour. Tetravalent materials such as germanium and silicon are semiconductors; carbon can be either an insulator (most types of diamond) or a semiconductor (graphite). Elements which have an odd number of electrons per atom are normally electrical conductors.

There is, however, one apparently unhappy disagreement with observation. This arises with the divalent elements such as magnesium, zinc, and mercury. Each of their atoms contributes an even number of electrons and so they should have completely filled electron bands. They should therefore be insulators, whereas in fact they are quite good conductors. This is explained away by assuming that for these metals the energy bands *overlap*,† and so

† Overlap can only occur in two- and three-dimensional systems.

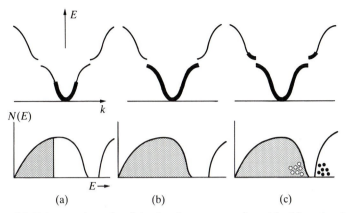

FIG. 8.5. E–k curves (upper) and density of states curves (lower) for (a) an electrical conductor; (b) an insulator (c) a semiconductor.

the electrons can pass from one band to the next without any thermal excitation (Fig. 8.6). The explanation of overlapping bands may sound like special pleading to make the theory fit the facts, but there is now sufficient experimental evidence to show that such band overlap does occur.

From Fig. 8.5(a) we see that if the electron distribution is such that the maximum occupied state lies near the centre of a band then the energy distribution is very similar to that predicted by the free-electron model and so the simple theory works quite well. It is only when the states are filled up to, or are very close to a band edge, that the free-electron model will probably be unsatisfactory and then the results of band theory must be used.

8.10. The relationship between energy and wave vector

The relationship between E and k near the top of, say, the first band needs to be explained in more detail. In quantum mechanics the velocity of a particle is given by the *group velocity* $d\omega/dk$ of the wave. Since $E = \hbar\omega$, this means that

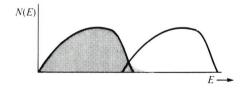

FIG. 8.6. Density of states curves for a metal with overlapping bands (e.g. a divalent metal).

the electron velocity is proportional to the slope of the $E \sim k$ curve. We therefore see that starting at $k = 0$ the velocity first increases with E. This can be thought of as being due to an increase in kinetic energy. But near the top of the band the slope of the curve decreases, so the velocity *decreases* in spite of the fact that the energy is still increasing. The reason for this is that the strength of the interaction of the electron with the periodic ionic potential has become significant. In classical terms one might think of this as giving rise to an increase in potential energy at the expense of some kinetic energy, just as a car slows down on encountering a hill. At the band edges dE/dk, and hence the electron velocity is zero. This, however, does not imply that the electron is stationary, but rather that motion in the $+$ and $-$ directions is equally probable, so that the expectation value of the velocity is zero. Nevertheless the expectation value of v^2, and hence of the kinetic energy, is non-zero.

8.11. The band theory in three dimensions; Brillouin zones

We have so far discussed the results of the band theory for a one-dimensional k. Since, however, k is a vector the edges of the bands should be described by the boundaries of a three-dimensional shape in k-space. Such a figure is called a *Brillouin zone*. Its shape is determined by the geometrical structure of the *translation* lattice (see section 1.2, p. 2), and its dimensions depend on the lattice constant a.

The Brillouin zones are constructed by taking any reciprocal lattice point as centre and then drawing perpendicular bisectors between that point and all surrounding reciprocal lattice points. The figures thereby produced are the Brillouin zones. It will be recalled (section 2.8) that in the construction for diffraction the incident wave vector terminated on the perpendicular bisector of a pair of reciprocal lattice points and hence any wave vector which ends at the Brillouin zone is one which would be diffracted.

A simple cubic lattice has a simple cubic Brillouin zone with a side of $2\pi/a$. Outside the first zone there will be a second zone which is bounded by those k values which are at the second energy gap, and so on for the higher zones. The first zone for a f.c.c. lattice is shown in Fig. 8.7.

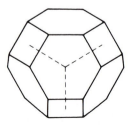

Fig. 8.7. The first Brillouin zone for a face-centred-cubic lattice.

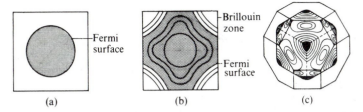

FIG. 8.8. The Fermi surface within the Brillouin zone for (a) a simple cubic lattice with a small number of free electrons for which the Fermi surface is spherical and well removed from the zone boundaries. This is the free-electron model. (b) As (a) but with more electrons so that the Fermi surface is distorted and touches the zone boundaries in some regions. (c) The Fermi surface of copper which touches the hexagonal faces of the zone (A. B. Pippard (1957), *Phil. Trans.* A**250**, 325).

Within these zones we should draw surfaces of constant energy. For small values of k these will be spherical and the electrons will fill up states within a sphere centred at the middle of the zone, just as in the free-electron case of Fig. 8.8(a). Towards the edge of the zone the constant energy surfaces become distorted, i.e. E depends on the *direction* as well as the magnitude of k (Fig. 8.8(b)).

8.12. The Fermi surface

The surface which encloses all the occupied states is called the *Fermi surface*. It should not be confused with the Brillouin zone which is purely a construction which depends on the lattice geometry. Free-electron-like behaviour occurs if the Fermi surface is well within the Brillouin zone. More complicated effects arise if they are close to or touch one another. Thus an experimental study of the shape of the Fermi surface and its proximity to the Brillouin zone is fundamental to a detailed understanding of the physics of electrons in metals. Such investigations, which are beyond the scope of this book, are usually made by investigating the magnetic or electrical behaviour as a function of an applied magnetic field. Fig. 8.8(c) shows the first Brillouin zone of a face-centred cubic lattice within which is drawn the Fermi surface of copper. It is seen that there is contact between the two on the hexagonal faces of the zone.

8.13. The many-body problem

In all the previous discussion, both for the free electron and the band theories, we have presented the results of calculations for the E versus k relationship for a system containing a *single* electron. We have then assumed that if more than one electron is present the first will occupy the lowest single electron energy state, the second the next state, and so on. In this we have implicitly assumed that the energy states themselves are unaffected by the fact that we have packed ~ 10^{29} electrons into a cubic metre, instead of just one! A more advanced study of electrons in metals must take account of the influence of all the electrons on the single electron states, although it turns out that, except at short distances, clouds of electrons act as if they 'screen' one electron from the influence of the others.

8.14. The experimental observation of energy bands; photo-induced electron emission

The shapes of the energy bands in metals were first deduced from soft X-ray spectra and a refinement of the technique gives very good resolution. However a very powerful method has been developed to determine the energies and the shapes of *all* states in an atom or ion with extremely high precision. This is the technique of photo- (or X-ray) induced electron emission (sometimes called ESCA†).

In these experiments the specimen is irradiated with ultraviolet or X-rays of precisely defined energy E_{rad}, which is sufficiently high to remove electrons

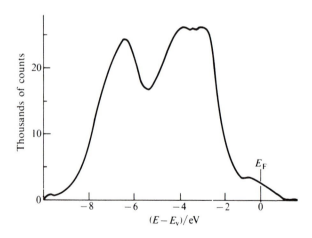

FIG. 8.9. The outer band of gold determined by ESCA measurements. This is composed of overlapping 5d and 6s bands. (After D. A. Shirley (1972), *Phys. Rev.* **B5**, 4709.)

† These are initials which stand for electron spectroscopy for chemical analysis.

from the crystal. The kinetic energy E_{kin} of these emitted electrons is measured in a very accurate β-ray spectrometer.†
If the energy to remove an electron from an atom or ion is E_{at} then

$$E_{at} = E_{rad} - E_{kin}$$

(neglecting certain correction terms). The technique can be used to measure very accurately both the sharp inner levels of an atom as well as the shape of the bands although since the electrons can only escape from within a depth of about 2 nm the data are only strictly relevant to the surface layers of atoms. Fig. 8.9 shows the overlapping 5d 6s bands for gold determined by ESCA experiments.

8.15. How do the band gaps arise?

Every author who sits down to write an introductory solid state physics text must approach the section on band theory with trepidation. The Schrödinger equation with a periodic potential (8.6) is quoted, and then all that is necessary is for a suitable wave function to be substituted in the equation and this will bring forth energy bands and energy gaps! Unfortunately a satisfactory treatment is not so simple, and a useful guess at the solution requires more intuition than is possessed by most people. And yet the whole idea of band gaps is so vital to our understanding of solids that it seems essential that some kind of reasonable justification for their existence should be given.
There is no doubt that the simplest qualitative explanation for the existence of bands and gaps is given by the tight binding theory which is described in section 8.19, but that is begging the question so far as eqn (8.6) is concerned.

8.16. Interference

First of all let us again emphasize the points we have already made in section 8.2. When *any* type of wave travels through and interacts with a periodic structure interference effects occur which for any arbitrary wave vector are destructive, and so the wave is undeviated. It is only for wave vectors which bear some special relationship to the lattice periodicity that there is the constructive interference which gives rise to diffraction.‡ We have already mentioned the analogy with X-rays, but we could equally well discuss radio-frequency filters which can be designed with stop and pass bands, optical filters, or mechanical anti-vibration mountings.

---

† Although the instrument is often referred to as a β-ray spectrometer the energies involved are much less (1–2 keV) than those measured in conventional β-ray work.
‡ Strictly speaking this is only for an infinitely large crystal. Why?

We therefore now demonstrate, with no attempt at rigour, how interference arises in (8.6) but first of all we must assemble a few mathematical facts. For simplicity we shall work in one dimension x.

8.17. Fourier representation and wave functions

(1) We may expand a function which has the periodicity a of the lattice as a Fourier series, the terms of which are periodic in multiples of a^{-1}, or more generally they are periodic in the reciprocal lattice vectors G. The periodic potential $V(x)$ may therefore be written as

$$V(x) = \sum_G V_G \exp{(iGx)} \tag{8.7}$$

and, if we wished, the coefficients V_G could be determined by conventional Fourier analysis (section 2.6, p. 24).

(2) The integral

$$\int_0^L \exp{(ikx)}\,dx = 0 \quad \text{if } k \gg L^{-1}, \tag{8.8}$$

because this is then just the integral over a large number of waves.

(3) If we have two complex periodic functions, $J(k)$ and $M(k')$ extending over a region of space from zero to L, then the integral

$$\int_0^L J(k)M(k')\,dx \qquad (k, k' \gg L^{-1}) \tag{8.9}$$

will be zero unless the product $J\,M$ is *not* periodic, i.e. for a non-zero result we need $k = -k'$. This is a sophisticated way of saying that two waves usually destructively interfere with one another.

(4) If the crystal extends over a length L the complete set of normalized free-electron wave functions is of the form $L^{-\frac{1}{2}} \exp{(ik_x x)}$, $L^{-\frac{1}{2}} \exp{(ik'_x x)}$... for the permitted values of k, k'... which satisfy the boundary conditions. Solutions of (8.6) can be represented by a suitable linear combination of these functions, e.g. $a_k L^{-\frac{1}{2}} \exp{(ik_x x)} + a_{k'} L^{-\frac{1}{2}} \exp{(ik'_x x)} + \cdots$.

(5) We now assume that the energy contribution ΔE of the potential term $V(x)$ in (8.6) to the total energy E is small (this is not necessarily always true) and that the main bulk of E is the same as is given by the free-electron theory, i.e. we assume that $V(x)$ adds a small *perturbation* to the energy. If this assumption is justified we can use *first order perturbation theory*† to calculate E.

† Don't panic, it's not that difficult! I realize that perturbation theory might be slightly beyond the general level of this book, but oversimplified methods of justifying the results of the Bloch equation are so tortuous as to be almost valueless. Even if the reader is not yet familiar with perturbation theory, all that is required is an acceptance of eqn (8.11) which, as any quantum mechanics text will show, is quite easy to derive.

8.18. Perturbation calculation

From (4) we must first choose suitable wave functions for the solution. Since in the free-electron case the $E \sim k$ relationship is doubly degenerate, i.e. both $+k$ and $-k$ correspond to the same value of E, this suggests that such a pair of wave functions might be a suitable solution for the equation. Therefore we shall use a wave function of the form

$$aL^{-\frac{1}{2}} \exp(-ikx) + bL^{-\frac{1}{2}} \exp(ikx). \tag{8.10}$$

First-order perturbation theory shows that if the wave function is a combination of two degenerate normalized wave functions ψ_{-k} and ψ_k then the extra energy ΔE due to a perturbation V is given by

$$\Delta E = \frac{1}{2}\left(\int_0^L \psi_{-k}^* V \psi_{-k}\, dx + \int_0^L \psi_k^* V \psi_k\, dx \right) + \frac{1}{2}\left\{\left(\int_0^L \psi_{-k}^* V \psi_{-k}\, dx \right. \right.$$
$$\left. \left. - \int_0^L \psi_k^* V \psi_k\, dx \right)^2 + 4\left(\int_0^L \psi_k^* V \psi_{-k}\, dx \right)^2 \right\}^{\frac{1}{2}}, \tag{8.11}$$

where in our case $\psi_{-k} = L^{-\frac{1}{2}} \exp(-ikx)$, $\psi_k = L^{-\frac{1}{2}} \exp(ikx)$, and $V = V(x)$, (note that the coefficients a and b are not included in ψ_{-k} and ψ_k).

We now expand $V(x)$ as in (8.7). The first term in (8.11) becomes

$$\int_0^L \psi_{-k}^* V \psi_{-k}\, dx = L^{-1} \int_0^L \exp(ikx) \sum_G V_G \exp(iGx) \exp(-ikx)\, dx$$

$$= L^{-1} \int_0^L \sum_G \exp(iGx)\, dx = \text{zero from (8.8)}$$

if we ignore any term for $G = 0$. (This would not contribute to any periodicity in V.) Similarly all the terms in (8.11) are zero except for the last one

$$\int_0^L \psi_k^* V \psi_{-k}\, dx = L^{-1} \int_0^L \exp(-ikx) \sum_G V_G \exp(iGx) \exp(-ikx)\, dx$$

$$= L^{-1} \int_0^L \exp(-2ikx) \sum_G V_G \exp(iGx)\, dx \tag{8.12}$$

$$= L^{-1} \int_0^L \sum_G V_G \exp\{i(G - 2k)x\}\, dx.$$

This is the vital term. If $G \neq 2k$ this integral will also be zero, but if any $G = 2k$ the exponential is unity, and so (8.12) is equal to V_G. Inserting this value of the integral into (8.11) we see that when $k = \frac{1}{2}G = +n\pi/a$ then the free-electron energy is changed by $\pm V_G$, i.e. there is an energy gap of $2V_G$ (twice the Fourier coefficient of the potential energy term corresponding to G).

This theory can be extended to show how the $E \sim k$ curve bends away from the free-electron shape for k values which are less than $\frac{1}{2}G$ (as in Fig. 8.4), and it can also be shown that at $k = \frac{1}{2}G$ the coefficients a and b in (8.10) have the relationship $a = b$ for $\Delta E = +V_G$, so that the wave function is a sine wave, and $a = -b$ for $\Delta E = -V_G$ when the wave function will be a cosine wave.

8.19. The tight-binding approximation

In the band theory we began with the assumption that the crystal is composed of positive ions which are permeated by quasi-free electrons and we then followed the consequences of this model. There is, however, a completely different way of treating the problem which gives further insight to the formation of bands.

In this alternative approach it is assumed that the crystal is formed by bringing individual atoms close together. When the atoms are well separated, so that each is an isolated system, the equivalent energy levels on each atom will all be exactly the same (Fig. 8.10(a)), but when the atoms are brought sufficiently close together so that there can be interaction between one atom and its neighbour they must be considered as a single system. The original equivalent energy states must now shift relatively to one another so that the Pauli exclusion principle is not violated. The closer the atoms are together, the greater will be this shift. It will clearly be more significant for the electrons

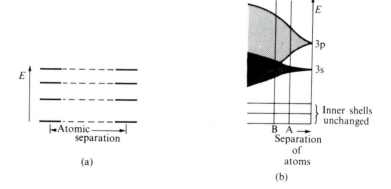

FIG. 8.10. The tight-binding approximation. (a) When atoms are well separated their levels are sharp. (b) Energy levels as a function of atomic separation showing how they broaden (at A) and overlap (at B) as the atoms are brought closer together. This demonstrates the formation of energy bands, band gaps, and overlapping bands.

in outer orbits. If there are N_i atoms, each electron state for a single atom will spread into N_i levels. This is shown in Fig. 8.10(b). Thus the 3s outer level of sodium, which can contain two electrons (although there is only one present in sodium) will spread into a band containing $2N_i$ discrete levels. Above this the 3p states (unoccupied in sodium) will also spread into another band. The breadth of the bands and the energy gaps between them will depend on the particular electron states of the atom which are involved, as well as on the atomic separation.

This treatment is called the tight-binding approximation because the outer electrons are assumed initially to be closely associated with their atoms. It shows that (1) the presence of band gaps or band overlap depends very much on the particular crystal and (2) since the bands are formed from the spreading of different electronic levels, the shape of a higher band is not necessarily the same as that of the lower band, nor need the number of permitted states in each be the same. Consequence (2) is of particular importance in band calculations for semiconductors. Neither (1) or (2) can be deduced directly from the quasi-free electron band theory.

PROBLEMS

8.1. The Fermi surface of a metal touches part of the Brillouin zone as, for example, do the outer curves of Fig. 8.8(b). Discuss how the electron distribution would shift if an electric field were applied in a horizontal direction. What effect would this type of Fermi surface have on the electrical conductivity?

8.2. At low temperatures the electrical resistivity produced by a given number of impurity ions in a metal is proportional to $(\Delta Z)^2$, where ΔZ is the difference in the valency of the impurity ions and that of the host metal. A sample of high-purity copper has a residual resistivity of $10^{-10}\,\Omega\,m$. When alloyed with 0·1 atomic per cent of Cd^{2+} the resistivity increases to $5 \times 10^{-10}\,\Omega\,m$. Estimate the resistivity of alloys of copper alloyed with 0·1 atomic per cent of (a) In^{3+} and (b) Sn^{4+}.

8.3. The thermal conductivity of a sample of germanium at 300 K is 80 W m^{-1} K^{-1}, and its electrical resistivity is $10^{-2}\,\Omega\,m$. Calculate the ratio of the electronic heat conductivity to the lattice heat conductivity of the specimen.

8.4. In a simple cubic quasi-free electron metal the spherical Fermi surface just touches the first Brillouin zone. Calculate the number of conduction electrons per atom.

8.5. Show that the number of different k-states in the first Brillouin zone of a simple cubic lattice is equal to the number of lattice sites.

8.6. Derive the relationship between the phase and the group velocities of a free-electron wave.

8.7. The Fermi energy of aluminium is 12 eV and its electrical resistivity at 300 K is $3 \times 10^{-8} \, \Omega$ m. Calculate the mean free path of the conduction electrons and their mean drift velocity in a field of 1000 V m^{-1}. (For Al, atomic weight = 27, density = 2700 kg m^{-3}.)

8.8. The periodic potential in a one-dimensional lattice of spacing a can be approximated by a box-shaped waveform which has a value at each atom of $-V$ and which changes to zero at a distance $0\cdot 1a$ on either side of each lattice point. Estimate the width of the first energy gap in the electron energy spectrum.

9. Semiconductors

9.1. Introduction

IN the previous chapter we briefly mentioned the electrical properties of semiconductors. At 0 K these materials have a completely filled electron band separated by only a small energy gap (~ 1 eV or less) from an empty band. There is no electrical conduction because the electrons are unable to change their energy states in small electric fields. At higher temperatures, however, there is sufficient thermal activation for some electrons to be excited from the lower band (the valence band) to the upper one (the conduction band) as in Fig. 9.1. An external electric field can now influence the electron states in both bands and a current can flow. The higher the temperature, the more electrons will be excited to the conduction band, and there will therefore be a tendency for the conductivity to *increase* with temperature. In a metal the conductivity decreases with temperature (Fig. 8.2).

The most important semiconducting *elements* are germanium and silicon. In the free atom these have four electrons in a half-filled outer shell (in the s and p states). In the solid these are covalently bonded to four neighbours (section 1.16, p. 14), the s- and p-levels are mixed,† and they form a completely filled valence band. The separation of the conduction band is ~ 0·75 eV for Ge and ~ 1·1 eV for Si. Most of the early fundamental and technological development has been carried out on Ge and Si and our treatment will concentrate on them, but the search for new materials for devices has led to an intensive study of semiconducting compounds which have other band gaps and differently shaped bands. The most important of these are composed of a trivalent with a pentavalent atom such as GaAs, GaP, InSb, InAs, and InP.

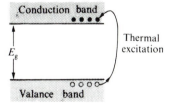

FIG. 9.1. The arrangement of electron bands in a semiconductor. Thermal excitation from the valence to the conduction bands occurs across the gap E_g. This results in mobile holes being left in the valence band and mobile electrons in the conduction band.

† As in the tight-binding approximation (section 8.19, p. 140).

9.2. Further consequences of the band model

In semiconductors we are particularly interested in the behaviour of those electrons which are near the band edges—those which are close to the top of the valence band and those near the bottom of the conduction band—because these are the electrons which are sufficiently close to vacant levels to be able to change their state when an electric field is applied. But it is just at the band edges that the effect of the crystal potential is greatest and so large deviations from the free-electron (parabolic) shape occur. The behaviour of those electrons which are near the top of the valence band is particularly significant and confusing.

9.3. The electron velocity and acceleration

In section 8.10 (p. 134) we explained that the slope of the $E \sim k$ curve is proportional to the group velocity v of the electron and from the argument in that section it is simple to show that

$$v = \hbar^{-1} \, dE/dk. \tag{9.1}$$

(More generally in vector notation $\boldsymbol{v} = \hbar^{-1} \, \mathrm{grad}_{\boldsymbol{k}} E$.) Thus just below the band edge v decreases even though E is still increasing (Fig. 9.2). This, as we already discussed is because E is the *total* energy; the potential energy increases at the expense of the kinetic energy. However, so far as the magnitude of the electric current is concerned, it is the velocity (dependent on the kinetic energy) which is important.

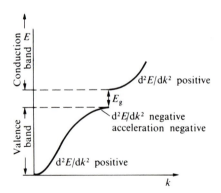

FIG. 9.2. The E–k relationship for positive k in one dimension for a semiconductor. The electron velocity is proportional to the slope of the curve and hence it tends to *decrease* as the top of the first band is approached.

From (9.1) we can derive a useful expression for the acceleration $\dot{v}$.

$$\dot{v} = \hbar^{-1}\left(\frac{d}{dt}\right)\left(\frac{dE}{dk}\right) = \hbar^{-1}\left(\frac{d}{dk}\right)\left(\frac{dk}{dt}\right)\left(\frac{dE}{dk}\right)$$

$$= \hbar^{-1}\left(\frac{dk}{dt}\right)\left(\frac{d^2E}{dk^2}\right). \tag{9.2}$$

9.4. The effective mass

If we apply an electric field $\mathscr{E}$ to the material there will be a force $-e\mathscr{E}$ on the electrons; the electrons will accelerate and their energy will change. The situation is complicated, however, because the electrons are influenced not only by the external field, which can be measured, but also by the internal field produced by the other electrons and by the periodic ionic potential, which is difficult to determine. But it is of course just these internal fields which determine the shape of the $E \sim k$ curve in the real crystal.

If the electron velocity is v, the rate of energy absorption from an external field $\mathscr{E}$ is force × velocity, i.e.

$$-e\mathscr{E}v = \frac{dE}{dt} = \left(\frac{dE}{dk}\right)\left(\frac{dk}{dt}\right). \tag{9.3}$$

We can substitute for v from (9.1)

$$-\left(\frac{e\mathscr{E}}{\hbar}\right)\left(\frac{dE}{dk}\right) = \left(\frac{dE}{dk}\right)\left(\frac{dk}{dt}\right)$$

and hence

$$-e\mathscr{E} = \hbar\left(\frac{dk}{dt}\right). \tag{9.4}$$

But from (9.2) we can substitute for dk/dt and obtain

$$-e\mathscr{E} = \left\{\frac{\hbar^2}{(d^2E/dk^2)}\right\}\dot{v}. \tag{9.5}$$

This is a relation between the force on the electron *due to the external field* and its acceleration within the crystal. Thus if we use Newton's law, force = mass × acceleration, and we calculate the electron motion by only taking account of the external field,[†] then $\hbar^2/(d^2E/dk^2)$ can be thought of as an effective mass m^* of the electron. Thus if E varies as k^2, m^* will be constant.[‡]

For an electron at the top of a band d^2E/dk^2, and hence the effective mass,

† This could be a *magnetic* field as well as an electric field.

‡ This will also occur at any band extremum since by a Taylor expansion ΔE, the energy from the band edge, should vary as k^2.

is *negative*, and so as we have already discussed, such an electron *decelerates* when a field is applied, momentum being exchanged with the lattice.

9.5. Holes

The conceptual difficulty associated with the negative effective mass of electrons at the top of a nearly filled band is simplified if we analyse the electron dynamics in terms of the small number of *vacant* states which are present there. These are called *holes* and we shall see that they can be considered to have an ordinary positive mass and a positive charge.

Let us first take a completely filled band containing n electrons which have velocities $v_1, v_2,..., v_n$. Since the band is full there are as many electrons with velocity $+v$ as there are with $-v$ and so the total current flow is zero and we may write

$$- |e| \sum_n v_n = 0,$$

where we have written the electron charge as $-|e|$ to make the presence of the negative sign completely unambiguous. Now let the lth electron be missing. There will be a net current equal to

$$- |e| \sum_{n \neq l} v_n,$$

but

$$- |e| \sum_{n \neq l} v_n + (- |e| v_l) = 0,$$

and so the current is equal to $+ |e| v_l$, i.e. the current could be considered to be due to a *positive* charge which has the velocity of the lth electron state. It is important to note that this hole has the *same* velocity as an electron would have in state l. It does *not* move (as is sometimes stated) with a velocity which is opposite to that of the lth electron. This is because the hole itself is on the 'emptier' side of the electron distribution (Fig. 9.3), and so its velocity is already in a direction which is opposite to the net electron current flow.

We must now determine the effective mass of the hole. Once again if the lth electron is missing, the total acceleration of charge may be written (using the same argument as for the velocity)

$$- |e| \sum_{n \neq l} \dot{v}_n = + |e| \dot{v}_l,$$

and from (9.5) we may substitute for $\dot{v}_l$ so that the net acceleration is

$$+ |e| \dot{v}_l = + |e| h^{-2} (\mathrm{d}^2 E / \mathrm{d}k^2)(- |e| \mathscr{E}). \tag{9.6}$$

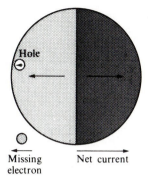

Missing   Net current
electron

FIG. 9.3. Holes in the valence band. If there is an electron missing from the band the hole has the same velocity as the missing electron. This is in a direction which is *opposite* to the net electron current.

But as we have seen $(\mathrm{d}^2 E/\mathrm{d}k^2)$ is negative at the top of a band and so we change the signs and write (9.6) as

$$+ |e| \dot{v}_l = + |e| h^{-2}(- \mathrm{d}^2 E/\mathrm{d}k^2)(+ |e| \mathscr{E}). \qquad (9.7)$$

The net acceleration is therefore the same as that produced by an external field acting on a positive charge which has an effective mass of $- h^2/(\mathrm{d}^2 E/\mathrm{d}k^2)$ which at the top of the band will be *positive*.

It is because holes have a positive effective mass that they can be considered as real positive charges and so the concept of hole conduction can be carried through in exactly the same way as for electron conduction. There would be no point in discussing the properties of a half-filled band in terms of the motion of the holes.

Energy-band diagrams are nearly always defined in terms of the energy of the *electrons*, but if we wish to discuss the energy of a hole then it must be measured downwards from the *top* of the valence band, because a hole which is well within the band has a higher energy than one close to the band gap.

9.6. Impurity levels

The presence of small carefully controlled amounts of certain impurities can dramatically affect the electrical properties of semiconductors and this has been crucial in the development of semiconducting devices.

If in a tetravalent semiconductor we introduce a pentavalent impurity atom there will be sufficient electrons to take up the four covalent bonds to all its neighbours, but the fifth electron is left fairly free. The electron is, of course, still bound to its own atom but theory shows that the radius of its orbit is much larger (several atomic radii) than that for the free isolated atom.

The ionization energy for this electron can be estimated by assuming that the effect of the other atoms (which are actually within the electron orbit) on this fifth electron can be taken into account by assuming that the orbit is filled with a medium which has the same dielectric constant as the semiconductor (this assumes the medium to be a continuum which unless the radius of the orbit is large is not really justified). Now the dielectric constants of semiconductors are high, about 16 for Ge and about 12 for Si and hence the force between this electron and its own nucleus is quite small.

The problem can be treated by modifying the Schrödinger equation for the hydrogen atom so that the electron, assumed to have an effective mass m^*, is embedded in a medium of dielectric constant ε. A straightforward analysis shows that the electron energy levels are those for the hydrogen atom multiplied by a factor $m^*/m\varepsilon^2$. Since the hydrogen energies are usually of the order of a few eV, the ionization energy for this special situation is $\sim 10^{-2}$ eV. The radius of the orbit is also correspondingly increased by a factor $m\varepsilon/m^*$ above that for hydrogen.

This low ionization energy means that a pentavalent atom can contribute an electron to the conduction band by thermal activation at quite low temperatures. Thus whilst at room temperature in a pure semiconductor there is very little excitation of carriers from the valence to the conduction bands, all the pentavalent impurities will be ionized at that temperature and their fifth electron will be available to participate in current flow. These atoms are called *donor* impurities. The energy level of the fifth electron is narrow and is usually just below the bottom of the conduction band (Fig. 9.4).

An analogous situation arises with trivalent impurities. They are one electron short for complete bonding to their tetravalent neighbours. This electron can be borrowed with very little energy—also about 10^{-2} eV—from a neighbouring atom thereby leaving a *hole* in the valence band. The presence of a trivalent atom therefore enables a host atom to be easily ionized and *hole* conduction can occur at room temperature (and below). These impurities are called *acceptor* impurities and the energy level is slightly above the top of the valence band as in Fig. 9.4.

Semiconductors which have a preponderance of *donor* atoms are called n-type materials (since the current carriers are negative) and those where the effect of acceptor atoms is dominant are called p-type specimens (the current carriers are positive—holes). If both types of impurity are present in similar concentrations there will be a tendency for one to cancel out the effect of the other. This is called *compensation* (Fig. 9.4). The addition of impurities is usually referred to as *doping*. If there is a large amount of doping the impurity levels can broaden and overlap the band edges.

These energy levels which are associated with the presence of tri- and pentavalent impurities are called *shallow* states. Other impurity atoms are more difficult to ionize and they have *deep* states which are well inside the energy

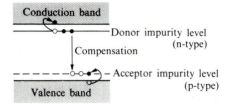

FIG. 9.4. Donor and acceptor levels in a semiconductor are within the band gap very close to the conduction and valence band edges respectively. If both types of level are present the donors will ionize the acceptors and reduce the effective numbers of both. This is called *compensation*.

gap. Whilst these states are important for certain properties we do not need to take account of them in our description of the basic behaviour of semiconductors.

For group IV semiconductors the most important acceptor impurity is B (also Ga, In, Tl) and the most important donor impurity is P (also As, Sb, Bi).

The influence of these impurities is very significant. Fig. 9.5 indicates how for n-type material they all ionized well below room temperature giving rise to a saturation region where the number of current carriers remains constant. Only substantially above room temperature will the thermal energy be sufficient for large numbers of electrons to be excited across the energy gap from the valence to the conduction bands. The room temperature behaviour is therefore dominated by the impurity levels and this is called *extrinsic* behaviour. Any properties which are associated with the electrons which are excited across the gap are called *intrinsic*.

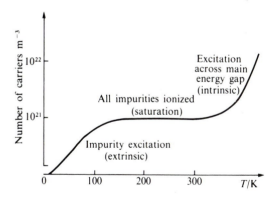

FIG. 9.5. The excitation of carriers in a semiconductor as a function of temperature. (Typical values for a specimen of doped germanium.)

Similar behaviour to Fig. 9.5 occurs for the excitation of holes in p-type material, although in the intrinsic region where both electrons and holes are being created, the electron current usually becomes dominant (because the mobility of electrons is normally higher than that of holes due to the different shapes of the conduction and the valence bands).

9.7. The determination of the number and charge of the current carriers

In order to evaluate a particular sample of a semiconductor we need to know the effective number of charge carriers which are excited at any particular temperature, the numbers of the various donor and acceptor states and the actual position of the impurity levels within the energy gap. The two standard measurements which are taken to determine these properties are the electrical resistivity and the Hall coefficient. The latter quantity is important because, as we shall show in the next section, it enables us to determine both the effective density N of the carriers and their sign. If we know N and the resistivity ρ then the mobility μ may be calculated from a relation which is really a definition of the conductivity σ

$$\sigma = 1/\rho = e\mu N, \tag{9.8}$$

where e is the electronic charge. The mobility is the average drift velocity of the carriers in unit electric field.

9.8. The Hall effect

If a current flows along a conductor which is in a magnetic field whose direction is perpendicular to the direction of current flow, then an e.m.f. (the Hall e.m.f.) is generated across the specimen in a direction which is perpendicular to the magnetic field, i.e. the current, the magnetic field, and the Hall e.m.f. are mutually perpendicular (Fig. 9.6). This phenomenon is called the transverse Hall effect. In general terms it is quite simple to see how this arises. The application of the magnetic field will cause the moving charges to deviate

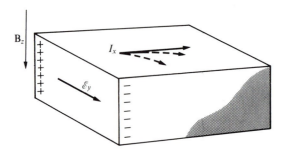

Fig. 9.6. The Hall effect (see text). I_x shows the direction of *electron* current.

from their motion along the specimen and their paths will curve towards the sides, where a space charge will accumulate. This will produce an electric field which will eventually be sufficient to counteract the effect of the magnetic field on the other moving charges and they will then transverse the length of the specimen without deviation. It is this electric field due to the space charge at the sides of the specimen which gives rise to the Hall e.m.f.

If we use the axes shown in the figure, then for electrons with a velocity v_x in a magnetic field B_z, the Hall field will be $\mathscr{E}_y$ in the direction indicated. If the charges are to be undeviated, the force due to $\mathscr{E}_y$ must be equal and opposite to that produced by B_z, i.e.

$$e\mathscr{E}_y = ev_x B_z = j_x B_z/N,$$

where $j_x = Nev_x$ is the current density. Now the Hall coefficient R_H is defined as

$$R_H = V_H z/(I_x B_z),$$

where I_x is the current through the specimen, V_H is the measured Hall e.m.f., and z is the thickness of the specimen in the z direction. Since $V_H = E_y y$ and $I_x = j_x yz$, we obtain

$$R_H = \mathscr{E}_y/(j_x B_z) = 1/(Ne). \tag{9.9}$$

It will be noted that R_H depends on the sign of e and the reader should verify that if $\mathscr{E}_y$ is in a certain direction for a flow of negative charges, then it will be in the opposite sense for the same current when it is produced by a flow of positive charges in the reverse direction.† If both electrons and holes are present together, the calculation is more complicated.

Since R_H is inversely proportional to the number of carriers, N, Hall measurements are very easy to make on semiconductors because N tends to be rather low, whereas for a metal the determination of R_H is more difficult. In the monovalent metals R_H is negative, which is consistent with our belief that the current is produced by a flow of negatively charged particles; the magnitude of R_H is then such that there is of the order of one moving charge per atom. In the more complicated metals, particularly those in which there is band overlap, R_H can be positive (e.g. in zinc and cadmium) and here it is assumed that most of the conduction occurs by the motion of positive holes.

9.9. The excitation of carriers; the Fermi energy of a pure semiconductor

In order to calculate the number of carriers N_e that will be excited into the conduction band at a temperature T we use the general formula,

---

† Confusion can arise because, since $\mathscr{E} = -\text{grad } V$, the direction from negative to positive surfaces is in the opposite direction to $\mathscr{E}_y$. If I_x is the direction of the flow of charges (whether they be electrons or holes) then I_x, $\mathscr{E}_y$, and B_z form a right-handed set of axes.

carriers excited = density of states × probability of occupation (sections 5.3 and 7.6, pp. 80, 111), i.e.

$$N_e = \int_{E_g}^{\infty} g(E) f_{FD}(E)\, dE, \tag{9.10}$$

where we have taken the top of the valence band as the zero of energy and E_g is the level of the bottom of the conduction band.

The Fermi–Dirac function $f_{FD} = [\exp\{(E - E_F)/(kT)\} + 1]^{-1}$ contains the normalizing parameter E_F which is called the Fermi energy and for *metals* this is equal to the energy of the highest occupied state at 0 K. It does not appreciably change from this value at higher temperatures (section 7.6, p. 111).

More generally, however, E_F is defined as the energy of a state for which the probability of occupation is $\frac{1}{2}$. Fig. 9.7(a) shows the situation at 0 K for an intrinsic semiconductor. The valence band is full and the conduction band is empty. f_{FD} is a step function which must therefore change its value from unity to zero within the energy gap and by symmetry it seems reasonable that E_F should be at $\frac{1}{2}E_g$. At higher temperatures (Fig. 9.7(b)) there are as many electrons excited into the conduction band as there are empty states (holes) in the valence band and so E_F remains at $\frac{1}{2}E_g$. This is only strictly true if the density of states $g(E)$ at the top of the valence band and at the bottom of the conduction band are exactly the same. If this is not the case E_F will vary slightly with T, but for most purposes E_F can be taken to be at the mid-point of the gap.

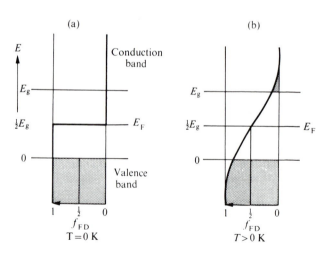

FIG. 9.7. The Fermi function f_{FD} for a pure semiconductor (a) at 0 K; (b) above 0 K. The Fermi energy remains approximately at the centre of the band gap.

9.10. The Fermi energy for doped specimens

The position of E_F for n- and p-type specimens is quite different from that for the intrinsic material. If we consider an n-type sample (containing no acceptor impurities) (Fig. 9.8(a)) at 0 K, all states up to and including the donor level are occupied and the conduction band is empty. E_F therefore lies somewhere between the donor level and the band edge. As the temperature is increased it remains in this region (Fig. 9.8(b)), falling slightly as more donors are ionized. It is only at very high temperatures when the intrinsic excitation is so high that there are almost as many holes excited as there are electrons, that the probability for 50 per cent occupation will be at $\frac{1}{2}E_g$ (Fig. 9.8(c)).

At room temperature, therefore, E_F for n-type material can be considered to be roughly at the donor level. A similar argument may be used to show that in a p-type material E_F is close to the acceptor level. This difference between the values of E_F for n- and p-type samples is extremely important and it provides the basic key for understanding the operation of p–n junction devices.

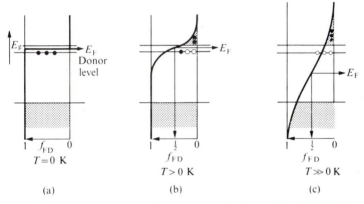

FIG. 9.8. The Fermi function f_{FD} for a n-type semiconductor (a) at 0 K; (b) above 0 K; (c) at much higher temperatures in the intrinsic region the function reverts to that in Fig. 9.7(b).

9.11. Calculation of the number of electrons and holes in a semiconductor

We now derive a very useful relationship for the product $N_e N_h$, where N_h is the number of holes in the valence band. If we assume that $(E - E_F) \geqslant kT$ (this is not always justified) then we may approximate the function f_{FD} by $\exp\{(E_F - E)/kT\}$ and so from (9.10) we may write

$$N_e = \int_{E_g}^{\infty} g_c(E) \exp\{(E_F - E)/kT\}\, dE,$$

where g_c is the density of states in the conduction band. This may be simplified by substituting $x = (E - E_g)/kT$, and hence

$$N_e = \exp\{(E_F - E_g)/kT\} \int_0^\infty g_c'(x, T) \exp(-x) \, dx.$$

In a similar manner we may substitute $y = -E/kT$ in an expression for N_h (the probability of occupation of a hole is $1 - f_{FD}$).

$$N_h = \exp\{-E_F/kT\} \int_{-\infty}^0 g_v'(y, T) \exp(-y) \, dy.$$

Therefore

$$N_e N_h = \exp(-E_g/kT) \times (\text{function of } T \text{ and the band shapes}). \quad (9.11)$$

It is important to note that this expression is independent of E_F and hence at a given temperature it is constant for *any* sample of a particular semiconductor. It is independent of the amount of doping or whether it is *n*- or *p*-type. Thus for a heavily doped *n*-type specimen the concentration of holes will be very small and similarly a heavily doped p-type material will not have many electrons. In a pure semiconductor (no donor or acceptor levels) N_e must equal N_h, and from (9.11) we may write

$$N_e = N_h = \text{constant} \times \exp\{-E_g/(2kT)\}. \quad (9.12)$$

It is interesting to note that the index of the exponent in (9.12) is $-E_g/(2kT)$ and *not* $-E_g/(kT)$, as might at first be thought for a process involving thermal excitation across an energy gap of E_g.

At room temperature the value of $N_e N_h$ is $\sim 10^{38} \, \text{m}^{-6}$ for Ge and $\sim 10^{33} \, \text{m}^{-6}$ for Si. Thus if there is no doping $N_e = N_h = \sim 3 \times 10^{16} \, \text{m}^{-3}$ for Si. This shows that in order to observe *intrinsic* behaviour in silicon at room temperature any ionized impurities must contribute considerably less than $3 \times 10^{16} \, \text{m}^{-3}$ carriers. The permissible content of such impurities must therefore be exceedingly small—of the order of 1 part in 10^{12} or 10^{13}.

9.12. The electrical conductivity of semiconductors; mobility of carriers

The conductivity of semiconductors is more complicated than that of metals because, besides the electron scattering processes being temperature dependent as in a metal, the actual number of current carriers N and their energy distribution also varies with temperature.

We already introduced the mobility μ in (9.8). This quantity is essentially a measure of the ease with which the carriers can move under the influence of an electric field, i.e. it is proportional to the inverse of the scattering probability. In metals we have seen that electron scattering is caused by phonons and by impurity atoms. In semiconductors the same two types of interaction apply, but their temperature-dependence is changed. This is because the energy distribution of the carriers in a semiconductor varies with tem perature, whereas in a metal it remains almost unchanged because the mean electron energy is so high. Since this dependence on the energy spectrum is complicated, we do not give any theoretical treatment of the problem.

In a semiconductor the effect of lattice vibrational scattering increases at higher temperatures (as it does in a metal) but the dependence is $T^{3/2}$. Impurity scattering, which was constant for a metal, tends to vary† as $T^{-3/2}$ (see Problem 9.13). Neither of these powers of T is as well established experimentally as are the corresponding relations for electron scattering in metals. The manner in which the two scattering mechanisms vary with T is shown in Fig. 9.9(a) from which it will be seen that the effective value of μ^{-1} has a minimum. If this is now multiplied by the inverse of the carrier concentration N^{-1} (from Fig. 9.5) we then find (Fig. 9.9(b), (c)) that the resistivity decreases very rapidly as the temperature is increased from 0 K, it passes through a minimum in the saturation region, and then at higher temperatures, in the intrinsic region, it falls rapidly with T. It will be appreciated that the details of the $\rho \sim T$ curve will be quite sensitive to the amount of doping in the specimen.

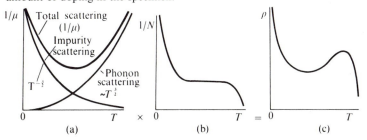

FIG. 9.9. Contributions to the resistivity of a semiconductor. (a) The scattering of the electrons (proportional to the inverse mobility) has impurity and phonon scattering contributions. (b) The inverse carrier concentration (from Fig. 9.5). (c) The resistivity is proportional to the product of (a) and (b).

9.13. Recombination and trapping

Electrons and holes have a finite lifetime after they are excited because as they drift they encounter other holes and electrons and then recombine. A direct recombination of a hole with an electron, however, is rather unlikely because they are both moving very rapidly and they would both have to be at the same place at the same time for recombination to occur.‡ A much more

† The $T^{-3/2}$ dependence is only for scattering by *charged* impurities in a semi-conductor.

‡ In Ge and Si there is a much more important reason why electron–hole recombination does not occur. Ge and Si are indirect gap materials (see section 9.17), and because a phonon would also have to be produced, the probability of recombina-tion is greatly reduced.

probable process is an indirect one which involves the 'trapping' of an electron (or a hole) by certain impurity atoms (which tend to have 'deep' electron levels, section 9.6, p. 148). They are thereby anchored in one place until a hole (or an electron) passes nearby and then the two can recombine. Trapping followed by recombination appears to be a much more effective mechanism than the direct recombination process. Even so, the carrier lifetime is relatively long. In Ge and Si it can be about 10^{-4} s. This long lifetime enables non-equilibrium distributions to exist and this is very important in the operation of the bipolar transistor.

Simple applications of semiconductors

9.14. Thermistors

The rapidly varying $\rho \sim T$ characteristic in certain temperature ranges makes some semiconductors very suitable for use as thermometers and as current controllers. These are both quite often called *thermistors*.

The doping of the material is usually arranged so that the thermistors are suitable for room temperature or biological temperature operation. The chip of semiconductor can be very small so that its presence causes little disturbance to a system. For example a thermistor can be mounted inside a fine capillary tube for temperature measurement in biological specimens.

It is also possible to obtain germanium sensors which are very sensitive at low temperatures down to the liquid helium region. These are probably the most stable secondary thermometers which are available at low temperatures.†

9.15. Magnetic-field measurement with a Hall probe

The large Hall e.m.f.s which are generated in quite modest magnetic fields can be used as the basis for a secondary standard for magnetic field determination. It is important, if the device is to be accurate, that the Hall e.m.f. should be closely proportional to the magnetic field and not too dependent on temperature. This may be achieved with certain semiconducting compounds such as InSb and InAs.

9.16. Infrared detection; photoconductivity

If radiation of angular frequency ω falls on a semiconductor and the quantum of energy $\hbar\omega$ is greater than the band gap E_g, then this radiation can be absorbed by a valence band electron which will thereby be excited into the conduction band. This absorption increases the number of carriers and hence

---

† A much cheaper thermometer which is often used is a carbon composition radio-resistor. Suitable carbon resistors ($\sim 100\ \Omega$ at room temperature) show a rapid increase in resistance below 4 K (10^3–$10^4\ \Omega$). It is not certain however that this variation is due to the properties of a bulk semiconductor. They have the disadvantage that on heating and recooling their calibration usually changes slightly.

it may be detected as an increase in the electrical conductivity. This is called photoconduction and it can be a very sensitive method of infrared detection. For the highest sensitivity the 'dark current' (i.e. with no radiation) is reduced by cooling the detector—usually with liquid nitrogen.

For the far infrared, when $\hbar\omega < E_g$, it is possible to detect the photoconductivity produced by the excitation of carriers from the donor or acceptor levels, although cooling to 4 K is often necessary to freeze carriers back into the donor or acceptor levels and so reduce the dark current.

9.17. Direct and indirect excitation

The condition for carrier excitation by electromagnetic radiation is not merely that $\hbar\omega \geqslant$ energy gap. In a crystalline material there is in addition the requirement that the wave vector must be conserved, i.e.

$$k_e + k_{\text{photon}} = k'_e, \tag{9.13}$$

where k_e and k'_e are the electron wave vectors before and after excitation respectively. Due to the very high velocity of light the photon wave vector is very small compared with k_e and k'_e, and so (9.13) reduces to $k_e = k'_e$. This means that for excitation across the narrowest part of the energy gap the minimum of the conduction band must have the same wave vector as the maximum of the valence band (Fig. 9.10(a)). This is called a *direct* transition.

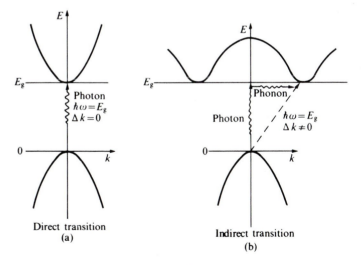

FIG. 9.10. Excitation across the band gap by photon absorption. (a) Direct process, no phonon is required. (b) Indirect process necessitates a phonon interaction in order that the wave vector is conserved.

Ge and Si have more complicated band structures and excitation just across the band gap often requires the additional interaction of a phonon in order that the wave vector conservation rule can be maintained (Fig. 9.10(b)). These are called *indirect* processes. For higher excitation energies direct transitions can occur.

A similar conservation condition applies for the *emission* of radiation which can occur when a free electron combines with a free hole and thereby drops back from the conduction to the valence band. This can limit the operation of light-emitting semiconducting devices (section 10.12).

9.18. Bolometers

Another method for detecting far infrared radiation is to use a suitably doped semiconductor as a *bolometer*. This consists of a small chip of material, usually cooled to 4 K, which warms up very slightly when it absorbs the radiation. The consequent change in the electrical resistance is then recorded as a measure of the infrared absorption. Bolometers can be exceedingly sensitive and they have been used to detect radiation which has a wavelength as long as 5 mm.

PROBLEMS

9.1. The Hall coefficient of aluminium is -0.3×10^{-10} V m A^{-1} T^{-1}. How many electrons per atom take part in electrical conduction? (For Al, atomic weight $= 27$, density $= 2700$ kg m^{-3}.)

9.2. Although the Fermi energy, E_F, for a simple intrinsic semiconductor is almost independent of temperature there is a small correction term. Show that $E_F = \frac{1}{2}(E_c + E_v) + aT$, where E_c and E_v are the energies of the states at the bottom of the conduction band and at the top of the valence band respectively and a is a constant. The densities of states in the bands may be assumed to be of the form $A(E - E_c)^{\frac{1}{2}}$ and $B(E_v - E)^{\frac{1}{2}}$ respectively.

9.3. A high-purity sample of germanium has intrinsic behaviour at 300 K. If λ^{-1} the threshold for continuous optical absorption is 5.5×10^5 m^{-1} estimate the temperature rise that will result in a 20 per cent increase in the conductivity.

9.4. A sample of silicon contains 10^{-4} atomic per cent of phosphorus donors which are all singly ionized at room temperature. The electron mobility is 0.15 m^2 V^{-1} s^{-1}. Calculate the extrinsic resistivity of the sample. (For Si, atomic weight $= 28$, density $= 2300$ kg m^{-3}.)

9.5. Eqn (9.11) shows that $N_e N_h$ is a constant at a given temperature. Give a physical explanation for the fact that when donor atoms are added to a sample to increase N_e, the number of holes N_h is correspondingly reduced.

9.6. A sample of n-type germanium contains 10^{23} ionized donors per cubic metre. Estimate the ratio at room temperature of the resistivity of this material to that of high-purity intrinsic germanium.

9.7. The effective electron mass in indium antimonide is ~ 0.01 electron mass. The dielectric constant is 17. Estimate the ionization energy of the donor atoms and the radius of the electron orbit.

9.8. A semiconductor is doped with 10^{22} donors and 5×10^{21} acceptors per cubic metre. The donor and acceptor levels are $10^{-2}\,\mathrm{eV}$ from their respective band edges. If the carrier mobility is $0.2\,\mathrm{m^2\,V^{-1}\,s^{-1}}$ estimate resistivity at 20 K. (See Fig. 9.4 for the effect of *compensation*.)

9.9. A Hall probe in the form of a cube of side 2 mm is being designed to measure a magnetic field. If the sensitivity required is 1 mV to detect $10^{-4}\,\mathrm{T}$, suggest a suitable carrier concentration for the probe. Assume a mobility of $0.5\,\mathrm{m^2\,V^{-1}\,s^{-1}}$.

9.10. A semiconductor chip of volume 1 mm^3 doped with 10^{20} donors per cubic metre is used to detect infrared at $10^4\,\mathrm{m^{-1}}$ by excitation across an extrinsic gap of $10^{-2}\,\mathrm{eV}$. If the chip is maintained at 4 K estimate the change in resistivity when it is illuminated with (and absorbs) $10^{-12}\,\mathrm{W}$, if the carrier lifetime is $10^{-6}\,\mathrm{s}$.

9.11. The $E \sim k$ relationship for the bottom of the conduction band of a semiconductor is of the form $E = Ak^2$, where $A = 5 \times 10^{-37}\,\mathrm{J\,m^2}$. Calculate the effective mass of the conduction electrons.

9.12. Show that if quasi-free carriers with an effective mass m^* are subjected to a constant magnetic field B then they will pursue circular paths at an angular frequency of Be/m^*. This is the principle of the technique of cyclotron resonance which enables the value of m^* to be determined in a very direct manner.

9.13 The density of states in a quasi-free electron band is proportional to $E^{1/2}$ (7.6) and the mean energy of a non-degenerate electron gas in the conduction band of a semiconductor is proportional to T. Thus the number of most readily available states into which an electron can be scattered is proportional to $T^{1/2}$. Use this to derive the $T^{3/2}$ and $T^{-3/2}$ dependences of the inverse mobilities due to phonon scattering and charged impurity scattering (section 9.12) respectively. Note that the number of phonons is proportional to T and that the deflection of charged particles (Rutherford scattering) is proportional to E^{-2}.

10. The physics of the semiconductor p–n junction

10.1. Equalization of the Fermi levels

THE junction which is formed when a piece of p-type semiconductor is in intimate contact with n-type material has very remarkable properties. It warrants considerable discussion since it is at the heart of many semiconducting devices.

The key to the understanding of the p-n junction rests on the following principles: (1) when two materials are in contact charge transfer occurs until their Fermi energies E_F are the same (section 7.14, p. 122), and (2) in n- and p-type materials E_F lies approximately at the donor and the acceptor levels respectively (section 9.10, p. 153).

Initially when n- and p-type materials are placed in contact the excess electrons on the n-type side can reduce their energy by diffusing into empty states in the p-type material, thereby charging it negatively (Fig. 10.1(a)). But the electric field which this produces will tend to oppose the diffusion of further electrons. This may be expressed in another way by saying that the energy levels on either side of the junction are displaced relative to one another (Fig. 10.1(b)). Eventually, a dynamical equilibrium will be established at a point where the relative shift of the levels is such that the transitions at any particular energy are equal on either side of the junction (Fig. 10.1(c)).

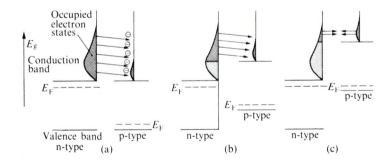

FIG. 10.1. Contact between n- and p-type materials. (a) Initially far more electrons diffuse from n to p than in the reverse direction. (b) The net diffusion, n to p, charges p negatively and there is a relative shift of the energy levels. (c) Finally dynamic equilibrium is achieved when the levels have shifted so that there are equal transitions of carriers at corresponding energies on either side of the junction.

This occurs when the E_F levels are coincident.† (This is called the principle of detailed balance.) Of course there will be another contribution to the process which will be provided by the hole diffusion from the p- to the n-type material and throughout this chapter, which we shall tend to discuss in terms of electrons, it should be borne in mind that there will be an analogous process produced by the holes. The effects due to these processes are additive.

10.2. The depletion layer

The initial surge of electrons from the n- to the p-sides of the junction will denude the n-type material of electrons in the region close to the junction and the only charges which will be left there will be the static positively charged donor *ions* (Fig. 10.2(a)). Similarly on the p-type side there will be a region containing only negatively charged acceptor ions. The junction region therefore consists of two *depletion layers* containing *fixed* equal but opposite charges (the proof that they are equal is given in the next section) and these can be considered to behave like a charged capacitance.

The amount of charge that needs to be transferred to equalize the Fermi levels is very small. Since in p- and n-type material the value of E_F differs by an amount roughly equal to the band gap (~ 1 eV) and the capacitance of, say, 1 mm cube of material is 10^{-1} pF, the charge necessary to raise the potential by 1 V is 10^{-13} C; this is less than 10^6 electrons.‡

10.3. Electrostatics of the depletion layer

The effective width and capacity of the depletion layer may be calculated quite easily by solving Poisson's equation.§

$$d^2 V/dx^2 = -\rho/\varepsilon\varepsilon_0, \tag{10.1}$$

where ρ is the charge density, which in the present case we shall write as ρ_a or ρ_d, due to the acceptor ions (negative) or the donor ions (positive) in the p- and n-type materials respectively. Note that we are here concerned with the *fixed* ionic charges and *not* with the mobile current carriers.

---

† A clear proof is given by A. van der Ziel (1968), *Solid state physical electronics* (2nd edn), pp. 51–3, Prentice-Hall, New York.

‡ This estimate, although of the right general order of magnitude, is only very rough for we have neglected the capacitance of the junction which may be several hundred picofarads.

§ The derivation of Poisson's equation will be found in any text on electricity and magnetism; e.g., B. I. Bleaney and B. Bleaney (1976), *Electricity and magnetism* (3rd edn), p. 32. Clarendon Press, Oxford.

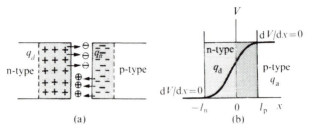

FIG. 10.2. (a) The depletion layer at a p–n junction. (b) Calculation of the width of the layer (see text).

If, as in Fig. 10.2(b) the junction is at $x = 0$ and the depletion layer extends to $-l_n$ and $+l_p$ into the n and p-type materials respectively, then we can integrate (10.1) and obtain for the n-type side

$$\varepsilon\varepsilon_0 \, dV/dx = -\rho_d x + \text{constant.} \quad (10.2)$$

Since $-l_n$ is at the boundary of the depletion layer the electric field (i.e. $-dV dx$) will be zero there. The constant in (10.2) is therefore $-\rho_d l_n$, and so

$$\varepsilon\varepsilon_0 (dV/dx)_{x=0} = -\rho_d l_n. \quad (10.3)$$

Similarly on the p-type side we may write

$$\varepsilon\varepsilon_0 (dV/dx)_{x=0} = +\rho_a l_p. \quad (10.4)$$

dV/dx must be the same on either side of the junction $(x = 0)$ and so we may equate (10.3) and (10.4). This yields

$$-\rho_d l_n = +\rho_a l_p. \quad (10.5)$$

Thus on either side of the junction the charges are equal and opposite. If N_d and N_a are the respective *numbers* of donor and acceptor atoms per unit volume then, taking account of the fact that ρ_d and ρ_a are of opposite sign, we have

$$N_d l_n = N_a l_p. \quad (10.6)$$

We therefore see that very heavy doping on one side will reduce the width of the depletion layer on that side.

If we integrate (10.2) between $x = -l_n$ and $x = 0$, we obtain

$$\varepsilon\varepsilon_0 (V_0 - V_n) = \tfrac{1}{2}\rho_d l_n^2 - \rho_d l_n^2 = -\tfrac{1}{2}\rho_d l_n^2 \quad (10.7)$$

and similarly on the p-type side

$$\varepsilon\varepsilon_0 (V_p - V_0) = +\tfrac{1}{2}\rho_a l_p^2.$$

The total potential across the junction, V, is therefore

$$V = \frac{1}{2\varepsilon\varepsilon_0}(-\rho_d l_n^2 + \rho_a l_p^2)$$

$$= \frac{-e}{2\varepsilon\varepsilon_0}(N_d l_n^2 + N_a l_p^2),$$

where e is the modulus of the electronic charge. Using (10.6) this may be written as

$$V = \frac{-e}{2\varepsilon\varepsilon_0}(N_d l_n^2)(1 + N_d/N_a).$$

If we now consider the depletion layers to act like a charged capacity then $-eN_d l_n$ will be the charge Q on one 'plate'.

$$\text{It follows that } V = \left\{\frac{Q^2}{2\varepsilon\varepsilon_0 e}\right\}\left(\frac{N_a + N_d}{N_a N_d}\right),$$

and so

$$Q = V^{\frac{1}{2}}\left\{2\varepsilon\varepsilon_0 e\left(\frac{N_a N_d}{N_a + N_d}\right)\right\}^{\frac{1}{2}}.$$

The capacity of the junction, $C = dQ/dV$, is therefore

$$C = \frac{1}{2}V^{-\frac{1}{2}}\left\{2\varepsilon\varepsilon_0 e\left(\frac{N_a N_d}{N_a + N_d}\right)\right\}^{\frac{1}{2}}. \tag{10.8}$$

If no external potential is applied, V is approximately equal to the gap potential. If an external potential is applied (this must be back-biased to avoid any current flow, see section 10.5) then V is equal to the sum of the gap and the applied potentials.

We therefore see that the p–n junction may be used as a capacitor which varies as $V^{-\frac{1}{2}}$. This is the basis of the variable capacity diode or 'varactor' which is used in frequency locking and frequency modulation circuits and also as a parametric amplifier.

If $V = 1\,V$, $N_a = N_d = 10^{22}\,m^{-3}$, and $\varepsilon = 10$, then C is about $2 \times 10^{-4}\,F\,m^{-2}$, or for $1\,mm^2$ of junction it is about $200\,pF$. The junction within most devices has an area which is much smaller than $1\,mm^2$, but it will still have a capacity of maybe up to $50\,pF$.

The width of one side of the depletion layer, say l_n, can be calculated from (10.7). If $(V_0 - V_{i_n}) = 0.5\,V$, $N_d = 10^{22}\,m^{-3}$, and $\varepsilon = 10$, then $l_n = 2.5 \times 10^{-7}\,m$, i.e. it is of the order of $1\,\mu m$.

10.4. Rectification at the p–n junction

Let us examine more closely the diffusion of carriers across the junction. We have already discussed in connection with Fig. 10.1(c) how the energy levels on either side of the junction are displaced so that there are equal numbers of carriers at similar energies on either side of the junction. The number of electrons n_p on the p-side is so small that they may all be considered to be in the lowest levels right at the bottom of their conduction band. The number which face them on the n-side will be proportional to $\exp(-E_g/kT)$, assuming Boltzmann statistics, where the energy of the electrons on the n-side is measured from the bottom of their conduction band (Fig. 10.3(a)) and E_g is the width of the energy gap. There will, of course, be continual diffusion of electrons across the junction, but once equilibrium has been established an equal number will travel in either direction. Now at any instant the number of electrons which diffuse from any level will be proportional to the population at that level and so we may write

$$C \exp(-E_g/kT) = An_p, \qquad (10.9)$$

where C and A are constants of proportionality for diffusion from the n- and p-sides respectively.

Let us now connect a battery which produces potential V across the junction in such a direction that the step between the conduction bands is lowered, i.e. the conventional positive terminal is connected to the p-side. This is called forward or positive biasing. The bottom of the conduction band on the p-side is now level with an energy† $E_g - eV$ on the n-side (Fig. 10.3(b)), and so there will be an imbalance in favour of those electrons crossing from n to p. However, since the junction is now part of a closed circuit no pile-up of electrons occurs to increase the step height again, because a current can flow.

The number of electrons which can now cross from n to p is increased to $C \exp\{-(E_g - eV)/(kT)\}$ since more of the n-side conduction band is now 'exposed' to the p-side. The number of electrons which can cross from p to n, however, cannot exceed An_p since this already takes account of *all* the available electrons on the p-side. There will therefore be a net transfer of electrons from n- to p-sides equal to

$$C \exp\{-(E_g - eV)/(kT)\} - An_p, \qquad (10.10)$$

and using (10.9) this can be written as

$$An_p[\exp\{eV/(kT)\} - 1]. \qquad (10.11)$$

---

† There is room for ambiguity here about the absolute sign of eV! All energies are positive for electrons and the sign of V is such that $E_g - eV$ is *less* than E_g!

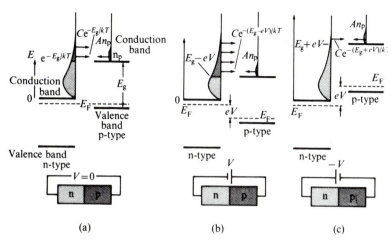

FIG. 10.3. Rectification at a p–n junction. (a) Zero bias (as in Fig. 10.1(c)). (b) Forward bias—the levels shift so that there is a net flow of electrons from n to p. (c) Negative bias.

It is clear that if V was reversed (reverse or negative biasing) the step between the n and the p sides would be *increased* to $E + eV$ and (10.11) would become $An_p\{\exp(-eV/kT) - 1\}$. This is negative, i.e. the concentration of electrons at the junction is *less* than in the main body of the p-type material (Fig. 10.3(c)) and there is a net transfer of electrons from p to n.

It should be noted that (10.11) gives the number of electrons in excess of the equilibrium number n_p which are present at any instant in the region very close to the junction on the p-type side, and this of course will produce an electron concentration gradient in the material. This expression, however, is *not* itself the current flow.†

10.5. The diffusion current

The electron concentration gradient which we have just mentioned will give rise to a diffusion of electrons either away from, or towards the junction, i.e. a current will flow. A calculation of the magnitude of this current is in general very difficult because of the many factors involved.

---

† Many simple treatments of the p–n junction derive (10.11) and then stop, suggesting that it is a calculation of the current through the junction, but this is only half the story. The general form of the expression for the current is indeed similar to (10.11) as we shall see, and all depends on the definition of A, but a proper understanding of the physics of the junction, however, requires a study of the effects of recombination. This is dealt with in the next section.

For forward biasing there will be an excess of electrons on the p-side of the junction. But because it is p-type material, as they diffuse away down the concentration gradient the electrons will meet many holes and they will tend to be trapped and recombine with them (section 9.13, p. 155). If recombination occurs, however, the material at that place has gained an electron, and in order to maintain charge neutrality there will be a flow of *holes* towards the junction. At any point in the material, therefore, the current consists of an electron diffusion current plus a hole drift current and their relative proportions will change as one moves away from the junction.

In order to solve this complicated problem we assume that very close to the junction, where the electron concentration gradient is a maximum, *all* the current is due to diffusion and, as we shall see, we can derive a useful expression for this diffusion current. Since, however, the current must be continuous around the circuit, this diffusion current will also be the value of the *total* current at all other parts of the material, even though in some places distant from the junction the hole drift current might be the dominant contribution (Fig. 10.4).

To calculate this diffusion current we define an electron–hole recombination time τ by

$$-\partial n/\partial t = (n - n_p)/\tau, \qquad (10.12)$$

where n is the electron density at any point and n_p is the equilibrium density in parts of the material which are distant from the junction. (10.12) just assumes that the net recombination rate is proportional to the excess number of electrons above the equilibrium concentration.

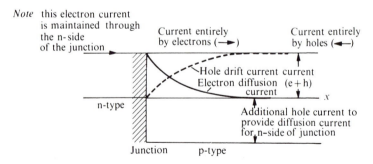

FIG. 10.4. The distribution of diffusion and drift currents on the p-side of a forward biased junction. Near the junction the current is mainly an electron diffusion current. Away from the junction this decreases and is compensated for by a drift current of holes so that the total current is maintained constant. Note that in addition to these currents there is also an extra hole current (shown at the bottom of the figure) which provides the hole diffusion current on the n-side of the junction.

The number of electrons which diffuse across unit area in the x direction is given by $-D\,\partial n/\partial x$, where D is the diffusion coefficient for the system. Let us consider what happens during unit time within a length δx of the p-type

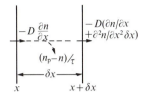

FIG. 10.5. Derivation of the diffusion equation (see text).

material (Fig. 10.5). We may write: number of electrons entering at x – number of electrons leaving at $x + \delta x$ = number of electrons recombining within δx, or, more formally,

$$-D(\partial n/\partial x) + D\{(\partial n/\partial x) + (\partial^2 n/\partial x^2)\delta x\} = -(\partial n/\partial t)\delta x = (n - n_p)\delta x/\tau$$

or

$$D(\partial^2 n/\partial x^2) = (n - n_p)/\tau. \tag{10.13}$$

This is a standard equation which has a solution of the form

$$n - n_p = \alpha \exp\{-x/(D\tau)^{\frac{1}{2}}\} + \beta \exp\{x/(D\tau)^{\frac{1}{2}}\}. \tag{10.14}$$

For a very long specimen it is clear that β is zero, otherwise the concentration of electrons would increase indefinitely as we move away from the junction. Hence

$$n = n_0 \exp(-x/L) + n_p, \tag{10.15}$$

where $L = (D\tau)^{\frac{1}{2}}$ is a *diffusion length* and n_0 is the excess population of electrons at $x = 0$, which has already been calculated (10.11).

The diffusion current at $x = 0$ is $-eD(\partial n/\partial x)$ and, from (10.11) and (10.15),

$$-eD\left(\frac{\partial n}{\partial x}\right)_{x=0} = \frac{eDn_0}{L} = \frac{eDAn_p}{L}\left\{\exp\left(\frac{eV}{kT}\right) - 1\right\}. \tag{10.16}$$

This is our final expression for the current.

If the potential is reversed it is evident that the current will flow in the opposite direction and the bracketed term in (10.16) would become

$$\{1 - \exp(-eV/kT)\}.$$

We therefore see that the p–n junction exerts a remarkable control on the current, and this depends on the direction of the applied potential (Fig. 10.6). If the p-side of the junction is made positive the current flows easily

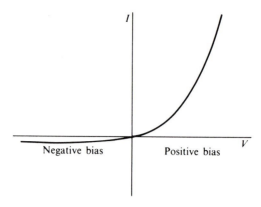

FIG. 10.6. The voltage–current characteristic for a p–n junction rectifier (see also Fig. 10.8).

and it increases rapidly as V is increased. Conversely if V is negative very little current flows and, for large negative values of V, it flattens off to a constant value.

We have pursued this argument in terms of the electrons in the p-type material. It is clear that there will be another contribution which will have a similar form to (10.16) which is due to the holes which diffuse away from the other side of the junction into the n-type material. The total current, it should be noted, will be the *sum* of these two contributions. There is no complete switch from electron to hole current at the junction! The electron current on the p-side which we have calculated will be maintained by a flow of electrons in the n-type material and similarly, the holes which are required for the current on the n-side will be provided by a hole current flowing from the p-side. This is indicated in Fig. 10.4.

10.6. The diffusion coefficient

An expression for the diffusion coefficient D may be derived by a simple argument.

If we have a piece of semiconductor in an electric field $\mathscr{E}$ and no current is allowed to flow (e.g. the material could be between the plates of a charged condenser) then in relation to carriers at the end $x = 0$, the potential at any other point will be $-\mathscr{E}x$ (Fig. 10.7). Thus at a temperature T the carrier density $n(x)$ at a point x will be proportional to $\exp\{q\mathscr{E}x/(kT)\}$, where q is the charge on a carrier. This would lead to a diffusion current $-Dq\{\partial n(x)/\partial x\}$, i.e. to $-q^2\mathscr{E}Dn(x)/kT$. But this current must be exactly countered by a drift current in the opposite direction due to the electric field, since the total current is zero. From the definition of mobility, μ, this drift current is equal to $\mu q\mathscr{E}n(x)$.

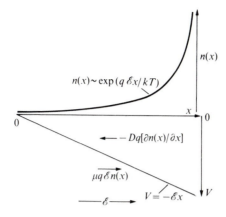

FIG. 10.7. Calculation of the diffusion coefficient (see text).

On equating these two currents we obtain the relation

$$D = \mu k T / q. \qquad (10.17)$$

This is sometimes called the Einstein relation, although his original derivation was for the equilibrium of particles in a gravitational field.

10.7. The effect of finite sample size

With the help of (10.17) we may calculate the diffusion length $L = (D\tau)^{\frac{1}{2}}$. A typical value of μ is $0.3 \, \text{m}^2\text{s}^{-1}\text{V}^{-1}$ and τ is around 10^{-4} s. These values yield a value for L of about 1 mm at room temperature. This is much larger than the width of the depletion layer (section 10.3) and indeed of most devices. We are therefore not usually justified in neglecting the second term in (10.14). For material of finite length Δ an adequate approximation is to substitute L by Δ in (10.15). This will also entail replacing L by Δ in the denominator of (10.16).

10.8. Minority-carrier devices

Our analysis has shown that the current which passes through the junction (10.16), is proportional to n_p on the p-side. It is controlled by the rate at which electrons can diffuse away from the junction into the p-type material and similarly by the rate of diffusion of holes in the n-type material, i.e. the control is by the *minority* carriers on either side of the junction. By suitable doping we can therefore arrange that most of the current will be carried either by the electrons or by the holes and this is very important in the design of bi-polar transistors.

10.9. The behaviour of the p–n junction in silicon

The current–voltage characteristic in Fig. 10.6 more or less describes a germanium p–n junction but it is important to note that a silicon junction does *not* behave exactly like this. Since nowadays most devices are made with silicon the difference in its behaviour should be discussed.

When a silicon junction is biased in the forward direction virtually no current flows until the potential reaches ∼0.6 V and beyond this voltage it increases rapidly (Fig. 10.8). This is probably due to the fact that since the

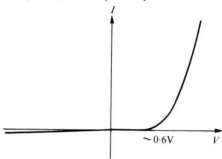

FIG. 10.8. The voltage–current characteristic for a silicon diode. In the forward direction no current flows until the bias is ∼0·6 V. The current for negative bias is actually extremely small. It would be of the order of nA if the positive current is of the order of mA.

product $N_e N_h$ is low for silicon (section 9.11, p. 154), the minority-carrier concentration is very small. This can provide only an extremely low current close to the origin and even with the exponential rise it takes until 0·6 V to build up to a reasonable value. Another contributory mechanism might be that due to the small minority concentration the electron traps on the p-side of the junction are not full. Hence immediately electrons cross from the n- to the p-side they are trapped (Fig. 10.9(a)). Only when V is sufficiently large that the step between the bottom of the conduction bands on either side of the junction is reduced and the traps can all be filled (Fig. 10.9(b)), are there any electrons left over to contribute to the conduction process. This behaviour at 0·6 V is very important in the design of circuits which use silicon transistors. The base–emitter potential must be set at about 0·6 V for satisfactory operation.

A brief discussion of the physics of some semiconductor-junction devices
10.10. The Zener and the avalanche diodes

If a *reverse* bias which is applied to a diode is increased then at a particular voltage, which depends on the doping of the elements, there is a very rapid increase in the current through the device (Fig. 10.10). This characteristic is particularly useful in voltage stabilization circuits.

There seem to be two effects which can produce the sudden increase in current.

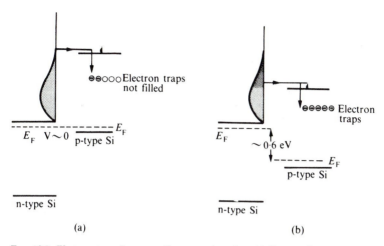

FIG. 10.9. Electron trapping at a silicon p–n junction. (a) For small positive bias, carriers fill empty traps and no current flows. (b) When the bias is ~0·6 V the traps are filled and the junction conducts.

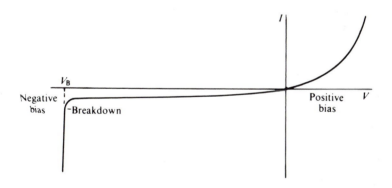

FIG. 10.10. The voltage–current characteristic of a Zener diode.

(1) The high electric field in the neighbourhood of the junction can accelerate the electrons to such a high velocity that they can, on impact, remove electrons from the atoms into the conduction band. These extra electrons can continue the process and a very rapid current build-up can occur.

(2) Another mechanism, which was first suggested by Zener, is that as the reverse potential is increased the junction region between the p- and n-type material becomes narrower until, at some particular value of the potential, quantum mechanical tunnelling can occur through the junction, say from the valence band on the p-side to the conduction band on the n-side.

The principles of both methods of operation are shown in Fig. 10.11. It has not yet been established in all circumstances whether it is (1) or (2) that is operative.

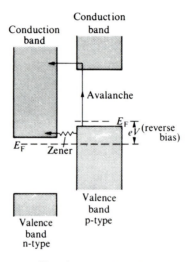

FIG. 10.11. Two processes of junction breakdown. In avalanche breakdown a high field excites an electron and gives it sufficient energy so that it can produce ionization when it interacts with other atoms. In Zener breakdown tunnelling occurs through the junction.

10.11. The tunnel (or Esaki) diode

We have up till now described the acceptor and donor states in doped specimens as being sharply defined levels. If, however, the materials are very heavily doped then these impurity levels are *broadened*. In the tunnel diode the doping is sufficiently great that these levels *overlap* the band edges so that the Fermi energy lies *within* the bands (Fig. 10.12(a)). The high doping also ensures that the p–n junction region is very narrow and that the minority-carrier concentration is small.

The device is operated with a *forward* potential V and because there are not many minority carriers the ordinary diode current is very low. Instead, carriers can *tunnel* through the junction region because the broadened

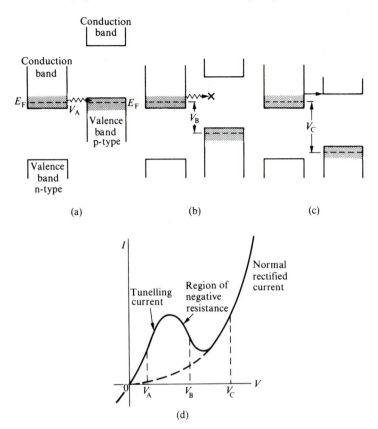

FIG. 10.12. The tunnel diode. (a) For a small positive potential electrons can tunnel through the junction (from n to p). (b) At a higher potential the levels on either side of the junction are shifted relatively to one another so that there are no permitted states on the p-side opposite the n conduction band and hence the current falls. (c) At a sufficiently large potential ordinary diode action occurs and the current rises again. (d) The current–voltage characteristic showing the three regions associated with (a), (b), and (c).

levels are roughly opposite one another (Fig. 10.12(a)). However, as V is increased each of the impurity levels starts to face the band gap on the other side of the junction and then tunnelling is no longer possible and the current starts to decrease (Fig. 10.12(b), (d)). At a sufficiently large value of V the potential step is so far reduced that ordinary diode action takes over and the current begins to increase again (Fig. 10.12(c)). The current–voltage curve (Fig. 10.12(d)) therefore has a region of *negative slope* (the device is said to have a negative differential resistance) and this may be used to advantage in the design of high-frequency amplifiers and oscillators. Since the existence of the negative-resistance region does not rely on thermal excitation, the tunnel diode operates satisfactorily at liquid-helium temperatures.

10.12. The light-emitting diode

In section 9.17 (p. 157) we explained that electromagnetic radiation can, under certain circumstances, be emitted when an electron and a hole recombine in a direct process. In order for this to be observed the population of the electrons in the conduction band must be increased above the thermal equilibrium value. There will then be more electrons dropping into the valence band *emitting* radiation than there are electrons being excited into the conduction band, thereby *absorbing* it. Such a population inversion, as it is called, is of course just what occurs near the junction of a forward biased diode and therefore with suitable materials light can be emitted.

As usual, things are not so simple in practice, because whilst, as we suggested in section 9.17, radiation could only be emitted from materials in which *direct* transitions can occur, the presence of impurity levels makes this restriction less rigid. Commercial light-emitting diodes are often made from GaP, which has an *indirect* band structure and in these devices radiation is emitted by transitions between *impurity* levels.

10.13. The bipolar transistor

Expositions of the operation of the transistor are legion (and sometimes erroneous), and this is not the place for a detailed discussion of their characteristics and behaviour. However, since we have worked through nearly all the physics which is required, we can hardly stop short at this point. Nevertheless, this brief discussion deals only with the basic principles.

A transistor is an amplifying device which is primarily used in circuits which are designed so that it acts as a voltage or as a power amplifier. It consists of three semiconducting regions arranged as a p–n–p or as an n–p–n sandwich. The two junctions which are thereby formed in the sandwich are biased, one in the forward and the other in the reverse direction. In this section we shall discuss the n–p–n arrangement (Fig. 10.13) but the general principles will also apply to the p–n–p device. The central slice of material is called the *base* and this is biased positively with respect to the *emitter* and negatively with respect to the *collector*.

The essential feature in the design is that the emitter is doped much more heavily n-type than the base is doped p-type and since at a given temperature $N_e N_h$ is a constant (eqn (9.11)) this means that there are far more (say a factor

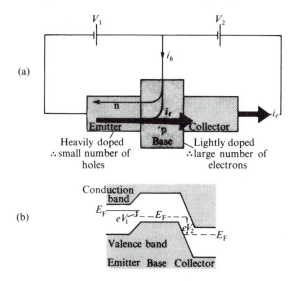

Fɪɢ. 10.13. The n–p–n transistor (common base circuit). (a) The emitter–base current is almost entirely due to electrons which diffuse across the thin base to the collector. This current can be controlled by making very small changes in V_1. (b) Relative disposition of the energy levels in the three sections of the transistor in a normal biasing scheme.

of ~ 10^2) minority carriers (electrons) in the base than there are holes in the emitter.†

Since it is the *minority* carriers which control the currents in the respective sections it follows that most of the current in the emitter–base circuit is an *electron* current. The base, however, is made very thin (~ 1 μm), and so this electron current can diffuse right through it with very little recombination. When it reaches the reverse-biased base–collector junction it is swept into the collector circuit by the large positive potential on the collector and it becomes the collector current.

The hole current across the base–emitter junction is usually so small that it plays no part in transistor action and indeed, if it were appreciable, a study of the rest of this section should make it clear that the transistor would not work very well. However, those electrons injected into the base from the emitter which do recombine in the base cause a small current of neutralizing holes to flow into the base from the external base lead. Any change in the base–

† Because it relies on both electrons and holes for its operation the transistor is called a *bipolar* transistor.

emitter potential produces a large change in the injected emitter electron current but a much smaller, though proportional change in the base current. Since most of these injected electrons flow on to the collector a small change in the base–emitter bias results in a large change in the collector current but only in a small change of base current, and so the input (base) circuit is not loaded.

Although the basic mechanism of transistor action is connected with a process yielding current gain† it should be noted that voltage amplification can be achieved because only very small voltages are required to change the emitter current. Since nearly all of this current becomes the collector current much larger voltage changes may be developed across a resistance of high value placed in series with the collector lead. (This will, of course, make the collector voltage change, but this hardly affects the collector current.)

There are of course many other aspects of transistor operation. It should be noted that the forward-biased emitter–base junction has a low impedance (a high current–voltage slope) whereas the reverse-biased base–collector junction has a high impedance and these characteristics can be used to advantage in power amplification. The circuit in Fig. 10.13 is called a common base connection. This is useful for giving a simple explanation but in practice it is more usual to use a common emitter (Fig. 10.14(a))—or more rarely, a common collector circuit (Fig. 10.14(b)) in both of which the input and output currents are separated.

The high-frequency performance of a transistor is limited by charge-storage in the base and by the internal capacity of the elements, the junctions and the leads, but these details, whilst important, take us beyond the fundamental principles of operation.

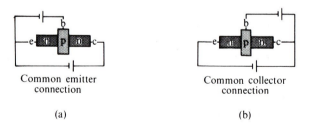

Common emitter
connection

Common collector
connection

(a)   (b)

FIG. 10.14. The transistor, (a) common emitter, (b) common collector, connections.

† This current gain is usually written as β or h_{fe}. It is usually greater than 10 and a typical value would be 150.

10.14. The field-effect transistor

The field-effect transistor (FET) is a device which overcomes one of the main disadvantages of the bipolar transistor—its relatively low input impedance. The basic construction is most easily described by reference to Fig. 10.15(a). A junction is made between thin strips of n- and p-type material. The p-type, say, which would then be called the gate, is reverse-biased with respect to the n-type, which itself has contacts at either end. These contacts are called the source and the drain and a potential of a few volts is maintained between them (the lower potential is connected to the source). A current therefore flows along the channel between the source and the drain.

The principle of operation is to control this channel current by varying the reverse potential on the gate. If this reverse bias is increased the depletion layer on the n-type side becomes wider (section 10.3) as more electrons are attracted across the p–n junction. The effective conducting channel between source and drain becomes narrower and so the current decreases.

The FET is *not* a bipolar nor a minority-carrier device. As we have described it, the current is determined by the flow of *electrons* in the n-type material; holes are not involved. It could be made, of course, to operate with the p-type material as the conducting channel and the n-type as the gate. Since the gate circuit passes virtually no current it has a very high input impedance—of the order of $10^9 \, \Omega$.

10.15 The current characteristic of the FET

The relationship between the channel current and the gate voltage could be derived, as in section 10.3, by using Poisson's equation to obtain an expression for the change in the width of the depletion layer. We have seen, however, that the depletion layers can act as a capacitor and it is in fact much simpler (and the result is in much better agreement with observation!) to treat the junction as a charged capacitor which anchors charges on its 'plates' and so prevents them from contributing to the current flow. The problem is slightly complicated by the fact that the potential between the gate and the channel varies along the channel and so the effective cross-sectional area of the conducting part of the channel is not constant.

Let distances along the channel be measured from the source at $x = 0$; the drain is at $x = L$. The potential at a point x is $V(x)$ (Fig. 10.15(b)). Initially we shall measure all potentials relative to the gate. We assume that the effective capacity per unit length of channel is a constant C.† Then the charge

---

† Reference to eqn (10.8) would suggest that a constant value of C is unjustified. The impurity concentration and profile at the junction of an FET, however, is complicated and it behaves much more as a constant capacity than one which varies with V.

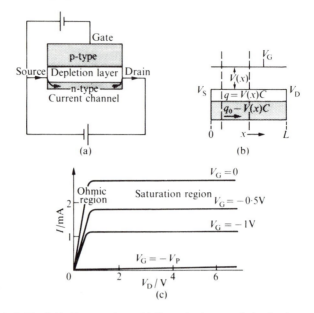

FIG. 10.15. The field-effect transistor. (a) General scheme and circuit; the potential on the gate alters the width of the depletion layer which thus controls the current from source to drain. (b) Calculation of the current (see text). (c) Variation of drain current as a function of drain voltage for various gate voltages. The plateau regime is the useful operating region.

per unit length on this capacity is $V(x)C$. This is the charge which is *not* now available for conduction. If q_0 is the charge per unit length when the potential is zero (i.e. the total possible charge which can contribute to current flow) then the charge, $q(x)$, now available for conduction will be given by

$$q(x) = q_0 - V(x)C.$$

Note that if $V(x) = q_0/C$, then *no* charge is available for conduction. This value of V is called the 'pinch-off' voltage V_p, and it occurs when the depletion layer has spread the full width of the channel. So we may now eliminate C by writing

$$q(x) = q_0\{1 - V(x)/V_p\}. \tag{10.18}$$

Now the conductance of the channel per unit length is $\mu q(x)$, where μ is the mobility of the carriers (9.8) and if the electric field is $\mathscr{E}$, then the electron current I is given by

$$I = -\mu q(x)\mathscr{E} = \mu q(x)\,dV/dx,$$

and so

$$\int I\,dx = \int \mu q(x)\,dV.$$ (10.19)

To integrate this between $x = 0$ and $x = L$ is quite simple because the current must be continuous throughout the length of the channel, i.e. I is independent of x, and so, after substituting (10.18) into (10.19), we have

$$IL = \mu q_0 \int_{V_S}^{V_D} \{1 - V(x)/V_p\}\,dV,$$

where V_S and V_D are the source and drain voltages respectively. A useful form of the solution is easily obtained by substituting $z = \{1 - V(x)/V_p\}$, and then

$$I = \frac{\mu q_0 V_p}{2L}\left\{\left(1 - \frac{V_S}{V_p}\right)^2 - \left(1 - \frac{V_D}{V_p}\right)^2\right\}.$$

If we now explicitly introduce the gate voltage V_G, this becomes

$$I = \frac{\mu q_0 V_p}{2L}\left\{\left(1 - \frac{V_S - V_G}{V_p}\right)^2 - \left(1 - \frac{V_D - V_G}{V_p}\right)^2\right\}.$$ (10.20)

This is the basic relationship between the current in the channel and the voltages at the various contacts. Eqn (10.20) may be rearranged to obtain simpler forms for various working conditions. For example by factorizing the difference of the squares and setting $V_S = 0$, we obtain

$$I = \frac{\mu q_0 V_D}{2L}\left\{2 - \frac{V_D - 2V_G}{V_p}\right\}.$$ (10.21)

The characteristics of an FET (for which (10.21) is very much an approximation) are shown in Fig. 10.15(c). For very small $V_D(\ll 2V_G)$ no current flows when V_G is equal to $-V_p$. If the gate bias V_G is made less negative the current is proportional to V_D provided that V_D is still small. The FET then behaves as a resistor whose value is controlled by V_G. The interesting region, however, is that where V_D is large. For a given V_G the current tends to saturate and it is almost independent of V_D. This is because an increase in V_D beyond V_p will 'pinch-off' part of the channel. (This is not predicted by (10.21) which shows a maximum in I with increasing V_D.) The current is maintained, however, since the region of the channel very near the drain drops nearly all the difference between V_D and V_p. In this short region of the channel the electrons are subjected to a strong electric field and so, although there are very few mobile electrons, their very high velocity allows them to transport all the current entering the region from the part of the channel near the source which is *not* pinched off.

Hence in the saturation range, changes in V_G can be used to produce variations in channel current which are more or less independent of the value of V_D, and this is the useful operating region of the device.

10.16 The MOSFET

A variant of the FET is the metal oxide-silicon FET or MOSFET. In this device the gate material is not a semiconductor but a thin metal film (aluminium), which is insulated from the silicon by a layer of SiO_2. The simplest form of MOSFET is shown in Fig. 10.16(a) but the more usual arrangement (Fig. 10.16(b)) has the source and drain connexions to the p-type channel made via small n-type regions. When the drain is biased positively with respect to the source, one of these junctions (which one?) will inhibit current flow in the channel. Conduction occurs when a positive potential is applied to the gate. This will create a very thin n-channel near the SiO_2 surface, which will provide a controllable conducting path between source and drain. Because the mobility of the electrons is so much higher than that of holes MOSFETs are usually made so that they operate with an n-type channel.

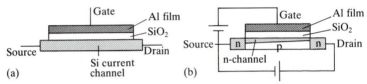

FIG 10.16.(a) Diagrammatic layout of a MOSFET. (b) The normal MOSFET has current connections of n material at the ends of the p-type conducting channel.

10.17 Two-dimensional systems and heterostructures

New devices are now being developed that consist of consecutive thin layers of different semiconductors, each having a carefully-controlled constant thickness—of the order of 10 nm. These are called *heterostructures*. While their detailed behaviour is beyond the scope of this book, the electronic properties of very thin films can be derived from a simple treatment, which illustrates some basic principles of quantum physics.

Let us first consider a very thin film of one semiconductor sandwiched between pieces of another whose band gap is larger than that of the film. When the Fermi energies equalize the situation will be as in Fig. 10.17.† An electron which has been thermally excited to the conduction band of an outer

---

† For simplicity Fig. 10.17 shows undoped materials. There are, however, band offsets at the junctions, which are fundamental properties of the two materials. These will ensure that similar potential wells occur with doped materials. See also Problem 10.12.

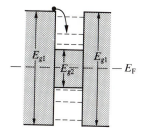

FIG 10.17. A thin film of a semiconductor with a band gap E_{g2} sandwiched between layers of another semiconductor with a larger energy gap, E_{g1}, gives rise to a quantum well. This will trap charges in sets of relatively widely-spaced quantum levels.

layer can reduce its energy by diffusing into the film and it will be trapped in this central potential well. What will be the distribution of energy states for electrons in this well?

This is the quantum mechanical problem of a particle in a very narrow box. If d, the thickness of the film, is in the z direction, the Hamiltonian in the well will be (from 7.1)

$$-(\hbar^2/2m)\partial^2\psi/\partial z^2) = E_z\psi$$

where E_z, the energy associated with motion of the electron across the film, will be given by $\hbar^2 k_z^2/(2m)$ (7.4). We can find approximate values by assuming an infinite potential barrier. The permitted values of k_z for standing wave solutions are $\pi n_z/d$, where the n_z are integers. If $d = 10$ nm, and the effective mass of the carriers is about 0·1 of an electron mass (which is typical for, say, GaAs) then the lowest value of E_z will be about 30 meV and the higher permitted z states will be separated by energies of a similar magnitude. Because these levels are discrete and widely-separated, the central film is called a *quantum well*. Due to the electron spin the states will be doubly degenerate.

Now the electron can also move in the x, y, plane of the film, and its total energy is $\hbar^2(k_x^2 + k_y^2 + k_z^2)/(2m)$, in which k_x and k_y can have very small values because motion in the plane is hardly constrained. The energy states associated with the x and y components of the velocity will be determined by the permitted values of $k_x^2 + k_y^2$. By following through an analysis similar to that used to obtain (7.6), it is quite simple to show that in the two dimensions x, y, the density of states is constant (see Problem 10.10). Thus the total energy of an electron in the well will be given by each value of E_z plus the permitted $(E_x + E_y)$ contributions. The final density of states, $g(E)$, will therefore have a step-like form; Fig. 10.18(a). It will start at the first permitted value of E_z and, because the density of states associated with $(E_x + E_y)$ is constant, $g(E)$ will have a constant value as E is increased, until

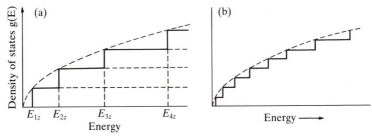

FIG 10.18. (a) The step-like density of states for the quantum well structure of Fig. 10.17. (b) As the film becomes thicker the steps become smaller and more closely-spaced and the density of states starts to approximate to the parabolic shape of a fully three-dimensional sample.

the second permitted value of E_z is reached. This value of E_z will then again be combined with all the permitted $(E_x + E_y)$ contributions and these states will be in addition to those formed with the first E_z value, and so on. Note in Fig. 10.18(b) how, if the film thickness is increased, the width of the steps becomes narrower and $g(E)$ will eventually revert to the dashed parabola that is typical of a fully three-dimensional system (Fig. 7.2).

Another way of illustrating the energy distribution is by the set of parabolas in Fig 10.19. The lowest parabola represents the $(E_x + E_z)$, k_x dependence for electrons in the lowest E_z level and successive higher parabolas have their minima at the higher values of E_z. (There will of course

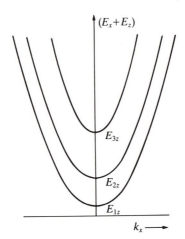

FIG 10.19. (a) Curves of $(E_x + E_z)$ versus k_x showing the separation of states into sub-bands. The minimum of each parabola is at one of the permitted values of E_z.

be a set of similar parabolas for the $(E_y + E_z)$, k_y states.) These curves are called *sub-bands*.

We have based our discussion on the electrons in the conduction band, but clearly a similar arrangement of widely-spaced quantum states will arise for the holes in the valence band.

The existence of these quantized states can be demonstrated experimentally by optical absorption. A photon of the correct energy can excite an electron from a valence band state to a conduction band state. Because in this transition wave vector must be conserved, $\Delta k = 0$, and strong excitations can only occur between a hole state and its corresponding state in the conduction band. As the frequency of the radiation is raised the absorption is observed to increase in steps which correspond to the onset of the higher E_z states.

The effects that we have described can be enhanced by having several films of the first semiconductor separated from one another by thick layers of the second. The arrangement then acts as several systems in parallel. If, however, the intervening layers are themselves *thin*, a new phenomenon can arise because charges from one quantum well can tunnel through an intervening layer to the next well; i.e. they behave as a *coupled* system. Providing that corresponding layers are all of precisely the same thickness we have introduced an additional periodicity, d, into the potential of the electrons in the wells. Such an arrangement is called a *superlattice*.† Since the z velocities and wave functions are now *not* confined to one well, the discrete values of E_z themselves become broadened (cf. the tight binding approximation, section 8.19, p.140). The situation is now very complicated but the end result is that just as a periodicity introduced band gaps into the free electron system at $\pm n\pi/a$ (Fig. 8.4, p.132) so this periodicity introduces gaps at $\pm n\pi/d$ into the bands of the semiconductor to form 'mini-bands' (Fig. 10.20). Because this periodicity and the strength of the interaction is under our control we now have the possibility of making semiconducting devices with a 'tailor-made' band gap, instead of having to rely on the gaps that nature has hitherto provided. This means that we can make devices which are sensitive to, or will operate at, energies in which we are especially interested.

10.18 Solar cells

The basic solar cell usually consists of a sandwich of a relatively thick (1 μm) undoped (and hence insulating) Si layer which has very thin (2 nm) n and p films on either side (Fig. 10.21). Just as in the conventional p–n junction (section 10.1) the Fermi energies throughout the device will equalize, thereby producing potential gradients across each junction. These will give rise to an electric field in the central insulating layer. When one face

---

† This usage is entirely different from that used in crystallography

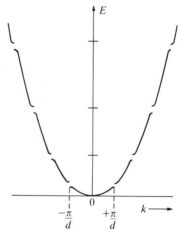

FIG 10.20. In a *superlattice* (see text) the additional periodicity breaks up each electron band into *mini-bands* with contain gaps at $\pm n\pi/d$.

of the sandwich is illuminated the photons penetrate to this layer, where they excite electron–hole pairs. The electric field sweeps the electrons and holes to the *n* and *p* regions respectively, and if the circuit between these films is completed a current will flow. The potential available to drive the current is usually about 0·5 V. Several devices can be connected in series to produce a higher potential.

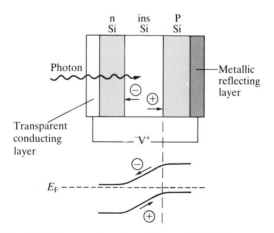

FIG 10.21. Simplified diagram of a solar cell (see text). The lower part of the figure shows how the electron energy levels change within the cell.

In practice the lower film is laid down on a metallic (Ag or Al) base, which not only acts as one connection but also serves to reflect back any light that reaches it so that it can be 'reused' in the cell. The top film is covered by a transparent conducting layer which acts as the other contact. Many complicated geometries have been devised in order to improve the cell efficiency. Solar cells are now used to power pocket calculators, watches, radios and other small devices, but this is a rapidly expanding technology and much larger cells are being developed.

PROBLEMS

10.1. Show that the value of the capacity of a p–n junction is the same as that of a parallel-plate condenser with a plate separation equal to the width of the depletion layers and filled with a dielectric with the same value of ε. A p–n junction of area 1 mm^2 is made of germanium with carrier densities 10^{21} m^{-3} acceptors and donors respectively. Estimate the width of the depletion layers and the capacity of the junction for zero applied voltage. (For Ge, energy gap $= 0.75$ eV, $\varepsilon = 16$.)

10.2. If electron tunnelling does not occur until the depletion layer is less than 10 nm, estimate the minimum doping which would be necessary in order that a germanium tunnel diode will operate ($\varepsilon = 16$).

10.3. Estimate the ratio of the forward and the reverse biased currents in a p–n junction diode when the applied voltage is 0.5 V.

10.4. The low-frequency performance of a n–p–n transistor at 300 K is seriously impaired unless the base region is very much thinner than 1 mm. If the electron mobility is 0.02 m^2 V^{-1} s^{-1} make an estimate of the electron–hole recombination time.

10.5. The mobility of electrons in a semiconductor is 0.2 m^2 V^{-1} s^{-1} and the recombination time for electrons is 10^{-4} s. A p–n diode is made from this material in which the p-type is doped with 10^{20} acceptors per cubic metre and the n-type is doped with 10^{23} donors per cubic metre. The junction area is 5 mm^2. Estimate the maximum current that could pass if the junction is forward biased with 0.5 V. (Assume $N_e N_h = 10^{33}$ m^{-6}.)

10.6. In an idealized diode the junction is plane. Show that the stored charge (i.e. the charge of the excess minority carriers) on one side of the junction is $q = (L^2/D)I = \tau I$ where I the diode current and D is the diffusion coefficient. L is the diffusion length and τ is the recombination time for minority carriers.

10.7. Calculate the Fermi energy, E_F, at 0 K for a quasi-free electron system with a charge density of 10^{20} m^{-3}. How will E_F vary as the temperature is increased? Compute the value of E_F by integrating the density of states $\times f_{FD}$ (see section 7.6), using trial and error if necessary.

10.8. A semiconductor which has a cubic lattice has the maximum of the valence band at $k = 0$ and the minimum of the conduction band for wave vectors which fall halfway along the $\langle 111 \rangle$ axes of the unit cell of the reciprocal lattice. The minimum band gap is 2 eV. A photon with energy equal to this minimum gap energy is emitted. Explain what other process must occur.

10.9 If V_p, the pinch-off voltage of a FET is 25 V, use (10.21) to plot curves of I against V_D for various values of V_G and determine the useful operating range. (Assume $\mu q_0/2L = 10^{-2}\,\Omega^{-1}$.)

10.10 Show that in a two-dimensional electron system of area A, the total number of s-states up to an energy E is $AmE/(\pi\hbar^2)$. Hence show that if the effective mass of the charges is unity the density of states is $2\cdot6 \times 10^{37}\,\mathrm{m}^{-2}$.

10.11 Calculate the energies of the first two z-states of an infinitely deep square quantum well whose z-dimension is 10 nm, if the effective mass of the carriers is $0\cdot1$. What carrier density (per m^{-2}) will be necessary in order that the electrons just start to populate the second state?

10.12 A film of GaAs (band gap $= 1\cdot5$ eV) is sandwiched between layers of $Ga_{0\cdot7}Al_{0\cdot3}As$ (band gap $= 1\cdot9$ eV). If the effective mass of the carriers is $0\cdot1$ and the film is 7 nm thick what is the maximum density of electrons which can be trapped in the well? Assume, as is usually the case, that the depth of the conduction band is 65 per cent of the difference in the band gap energies.

11. Paramagnetism

11.1. Introduction

THE magnetic properties of materials are a consequence of the fact that under certain circumstances some atoms and ions behave as elementary magnetic dipoles whose relative orientation is influenced by an external magnetic field. In this chapter we describe the behaviour of crystals which contain isolated (non-interacting) magnetic ions. In an external magnetic field such materials can acquire a magnetic moment which is proportional to the field. This is the phenomenon of paramagnetism which, although a rather small effect† at room temperature, needs to be described and understood before we proceed in the next chapter to describe the more well known and important technological property of ferromagnetism. The paramagnetism associated with the conduction electrons (not the ions) in a metal has been discussed in section 7.9 (p. 115).

11.2. The magnetic susceptibility

Magnetic properties are often discussed in terms of the susceptibility χ_v: this is defined‡ in terms of the magnetic moment m_v produced by a magnetic intensity H, as

$$\chi_v = m_v/H. \tag{11.1}$$

The susceptibility and the magnetic moment can be defined for a unit volume, unit mass, or a mole, and the reader is warned that authors do not always make clear which definition is being used. In this chapter we shall refer m_v and χ_v to a unit volume.

11.3. The origin of atomic magnetism; angular momentum

The motion of the electrons in the charge-clouds around the nucleus may be considered to have a similar magnetic effect to that of an electric current flowing in a closed loop of wire. The closed current loop behaves as if it

---

† We shall not discuss *diamagnetism* in which a magnetic moment is induced which *opposes* an applied magnetic field. This is a property of all atoms and ions but except in special circumstances (e.g. superconductivity) it does not materially influence the properties of solids. It is discussed in texts on electricity and magnetism, e.g. B. I. Bleaney and B. Bleaney (1976). *Electricity and magnetism* (3rd edn), p. 166. Clarendon Press, Oxford.

‡ Note that we follow the convention that the susceptibility is given by (11.1), but the energy of a dipole with a moment m_m is $-m_m \cdot B$. We call B the field and H the magnetic intensity.

possesses a magnetic moment, oriented normally to the plane of the loop, which is equal to the product of the current and the area of the loop, although of course if the time-averaged value of the electron velocity is zero (i.e. if on average it circulates as often in one sense around the nucleus as it does in the other) then the effective current is zero, as will also be its magnetic moment.

It is convenient when discussing the electron magnetic moment to relate it to the angular momentum of the electrum because, as the reader is probably aware, angular momentum is quantized in an atomic system and so the angular momentum quantum number is a useful way of designating not only the electronic state of an atom but also its magnetic moment.

Classically the angular momentum G is

$$G = m_e va, \qquad (11.2)$$

where v is the velocity and m_e is the mass of the electron,[†] which we assume to be in a circular orbit of radius a. This circulating electron is equivalent to a current $-ev/(2\pi a)$ and hence its magnetic moment m_m will be

$$m_m = -\pi a^2 ev/(2\pi a) = -eva/2,$$

and hence, from (11.2)

$$m_m = -eG/(2m_e). \qquad (11.3)$$

This is the classical relationship between the orbital angular momentum and its associated magnetic moment.

Besides its orbital motion an electron also spins on its own axis and the effect of this spinning charge is to give an additional contribution to the magnetic moment. However, the magnetic moment associated with this *spin* angular momentum is twice as large as that given by (11.3). This means that a knowledge of the *total* angular momentum is in general *not* sufficient to enable us to calculate its magnetic moment. We must know the proportions of the orbital and the spin components.

11.4. Quantum numbers

To discuss magnetic behaviour even qualitatively requires some familiarity with the manner in which the orbital and spin moments of the individual electrons in an atom interact and combine. We shall summarize the points which are important for our purpose with no attempt at justification. The motion of the electrons about a nucleus can be characterized by four quantum

---

† In this chapter we use m_e for the mass of the electron, whereas in the rest of the book we use m, which we here reserve for the magnetic quantum number. e as previously is the *modulus* of the electronic charge.

numbers. They are illustrated in Fig. 11.1. The energy is determined largely by the main quantum number n; this number is integral and $n = 1$ relates to the smallest value of the energy. The orbital angular momentum of an electron for a given n is characterized by l and this can have integral values which run from 0 to $(n-1)$. Associated with the angular momentum is an orbital magnetic dipole moment, as in (11.3), and in a magnetic field B this dipole can orient itself in a limited number of positions relative to B. The energy of the dipole will depend on its orientation and this is determined by the values of a magnetic quantum number m; this is also integral and it can have values $-l, ...0, ..., +l$. In addition to n, l, and m, the electron has a spin quantum number, s, for which there are only two possible values, $\pm\frac{1}{2}$.

The electrons will tend to occupy states of the lowest energy compatible with the Pauli exclusion principle, i.e. only one electron can be present in a state characterized by a given n, l, m, and s. Since s can only have the two values $\pm\frac{1}{2}$, we can also express this restriction by saying that not more than *two* electrons can have the same values of n, l, and m.

If all the states for a given n, l are occupied, their resultant spin dipole moment will be zero because there will be as many electrons with $s = +\frac{1}{2}$ as with $s = -\frac{1}{2}$. They will also have zero orbital dipole moment in the presence of a field because for each electron in a state $+m$ there will be another in a state $-m$. Such an assembly of electrons forms a *closed shell*.

The only electrons which will contribute to the dipole moment will therefore be those which are in incomplete shells. In general these will be the so-called valence electrons. Now the alignment of electron spins within an atom and

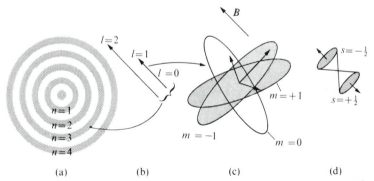

FIG. 11.1. The scheme of quantum numbers for an electron in an atom. (a) The main groupings of electron clouds are designated by n. (b) Within any n the angular momentum quantum number l can be $n-1$, $n-2$, ...0. (c) In an applied magnetic field each particular angular momentum l has a set of special orientations with respect to the field. These are such that the components m in the direction of the field are $m = l$, $l-1$, ... $-l$. (d) Each electron also has a spin angular momentum which can have either of two orientations in a magnetic field, designated by $s = \pm\frac{1}{2}$.

between neighbouring atoms is governed by a special type of interaction known as the *exchange interaction* (section 12.3 p. 201). The exchange interaction between the electrons in an incomplete shell is such that the energy is a minimum when the electrons line up with their spins parallel (provided that the Pauli principle is not violated). Since most atoms have an incomplete shell they will thus also have a dipole moment. It should be emphasized, however, that this is only true for *free* atoms and not for molecules. The reason for this is that the exchange energy of electrons of *neighbouring* atoms is usually a minimum when their spins are antiparallel and hence the total dipole moment of the molecule will be zero. In ionic crystals the outer electrons of one atom are transferred to complete the shell of its neighbour (section 1.15, p.13) so that both ions have closed shells of electrons around them and these also have no dipole moment.

Hence magnetic effects (apart from diamagnetism, footnote to section 11.1, p. 187) will usually only occur in solids when the atom has an incomplete shell in addition to that occupied by the valence electrons. An inspection of a table of the electronic configuration of the elements will show that there are five groups in which this occurs. These are the iron group, with an incomplete 3d shell, the palladium group (incomplete 4d), the lanthanides or rare earths (incomplete 4f), the platinum group (incomplete 5d), and the actinides (incomplete 5f). Almost all experimental work and practical applications have been concerned with salts of the iron or of the rare-earth groups and it is on these that we shall concentrate.

11.5. Coupling of the orbital and the spin angular momenta

The total angular momentum of the electrons in an atom is the vector sum of the spin and the orbital components and it is this resultant angular momentum which is quantized. If an atom possesses only one electron this raises no special problem, but if there is more than one electron, there are two ways in which a total angular momentum ground state of the system may be defined. The most common coupling scheme is one in which the total orbital angular momentum of all the electrons is combined with the total spin angular momentum. This is called $L-S$, or Russell–Saunders coupling. In the other scheme (which is more applicable to heavy atoms) the individual orbital and spin angular momenta of each electron are first coupled and then these individual resultants are combined together. This is called $j-j$ coupling and we shall not discuss it any further.

11.6. Russell-Saunders coupling; Hund's rules

The ground state of an atom or ion is determined by combining the quantum numbers of the individual electrons in accordance with a set of instructions known as Hund's rules. These yield a composite quantum number J. The rules which result from a consideration of the energies of the coupling mechanisms in descending order (spin–spin, orbit–orbit, total spin–total orbit), are as follows:

(1) the spins of the electrons are arranged so that as many of them as possible are parallel to each other without violating the Pauli principle (i.e. only two electrons to each value of m). Taking $s = +\frac{1}{2}$ or $-\frac{1}{2}$ for each electron, depending on the direction of spin, $\sum s$ is calculated. This sum is designated by S, the combined spin angular momentum.

(2) The electrons with the spins assigned as in (1) are divided amongst the possible values of m so that $\sum m$ is a maximum. This sum is denoted by L, the combined orbital angular momentum.

(3) The states of the atom or ion are characterized by a quantum number J which runs in integral steps from $L - S$ to $L + S$. The ground state, which is the only one with which we shall be concerned, is given by $J = L - S$ for a shell which is less than half full and by $J = L + S$ for a shell which is more than half full.

Thus Sm^{3+} which has five electrons in the f-shell ($l = 3$) will have them arranged with all spins parallel with the following m values

$$m \quad 3 \quad 2 \quad 1 \quad 0 \quad -1 \quad -2 \quad -3$$

$$s \quad \uparrow \quad \uparrow \quad \uparrow \quad \uparrow \quad \uparrow \ .$$

S will be $\sum s = \frac{5}{2}$ and L will be $\sum m = 5$. Since the shell is less than half full the ground state will have a value of J given by $L - S = \frac{5}{2}$.

11.7. The magnetic moment of an ion

Having established the angular momentum of an ion, as in the last section, we now need to calculate its magnetic moment. As we have previously mentioned, the relation between the angular momentum and the magnetic moment depends on the relative proportions of spin and orbital angular momentum. This information is contained in the Landé splitting factor g, which is defined by†

$$g = \frac{3J(J+1) + S(S+1) - L(L+1)}{2J(J+1)}, \tag{11.4}$$

and the magnetic dipole moment associated with a given value of J is then given by

$$-g\beta\{J(J+1)\}^{\frac{1}{2}}, \tag{11.5}$$

where β is the Bohr magneton.‡ For a system which has only spin (i.e. $L = 0$), $g = 2$.

† A derivation of (11.4) is given by B. I. Bleaney and B. Bleaney (1976). *Electricity and magnetism* (3rd edn), p. 444. Clarendon Press, Oxford.

‡ The Bohr magneton is the component of the magnetic moment, parallel to an applied field, of an orbiting electron with $l = 1$. This is equivalent to making $G = \hbar$ in (11.3). Thus $\beta = e\hbar/2m_e = 9.274 \times 10^{-24}\ JT^{-1}$.

11.8. Paramagnetism

When a magnetic dipole of moment m_m is placed in a magnetic field B it precesses about the field direction so that the direction of the moment maintains a constant angle with respect to an axis which is parallel to the field (Fig. 11.2). The energy of the dipole in the field is $-m_m \cdot B$, and unless there is a mechanism whereby this energy can be dissipated, e.g. by interaction with some other system, the precession will continue indefinitely. If we have a compass needle swinging in a field, the energy is dissipated by friction at the pivot and by air resistance; in the case of a magnetic ion in a crystal it interacts with the surrounding ions and so the magnetic energy can be dissipated by phonons. The precession angle will therefore be reduced and hence the moment will align itself as closely as possible with the field. It should, however, be emphasized that this can only occur if there is a mechanism for energy dissipation.

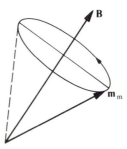

FIG. 11.2. On the application of a magnetic field B to an electron with magnetic moment, m_m the moment precesses about an axis which is parallel to the field.

On the atomic scale quantum mechanics shows that the moment has only a limited number $(2J + 1)$ of permitted orientations relative to field. These orientations are those in which the component of (11.5) parallel to the field is given by

$$m_J = -g\beta J, \; -g\beta(J-1), \; -g\beta(J-2), \; ..., \; +g\beta J. \tag{11.6}$$

The energy of the dipole in the field B can therefore have the values

$$-m_J B = g\beta J B, g\beta(J-1)B, ..., -g\beta J B \tag{11.7}$$

and no others.

If the dipole system is in thermal equilibrium with the crystal lattice at a temperature T then the relative numbers of dipoles with the various orientations and energies will be given by the Maxwell–Boltzmann distribution, $\exp\{-g\beta J B/kT\}$, $\exp\{-g\beta(J-1)B/kT\}...$, and so on.

At high temperatures the values of these exponentials are very similar to one another and so there will be almost as many ions whose magnetic moment is oriented antiparallel to the field as there are those which are parallel to it. The net magnetic moment of the material will therefore be very small and it will have a low susceptibility (11.1). If, however, the temperature is reduced (and the field is kept constant) then many more ions will have their moments parallel to B, thereby reducing their energy, and so the net total moment and the susceptibility will increase. This increase in the susceptibility as the temperature is reduced is the typical behaviour of a paramagnetic material.

11.9. Paramagnetic susceptibility; Curie's law

To calculate the susceptibility we first use (11.6) to write down the net value of the component of the total magnetic moment of the material which is parallel to the field B for a temperature T. This will be

$$m_v = \frac{N_v \sum_{m_J} m_J \exp(-m_J B/kT)}{\sum_{m_J} \exp(-m_J B/kT)} = \frac{N_v \sum_{-J}^{J} -g\beta J \exp(-g\beta JB/kT)}{\sum_{-J}^{J} \exp(-g\beta JB/kT)}, \quad (11.8)$$

where N_v is the number of magnetic ions per unit volume. The solution may be simplified provided that $g\beta JB \ll kT$ (except at very low temperatures this will be so).

If we set $x = g\beta B/kT$ and expand the exponential we then obtain

$$m_v = \frac{N_v g \beta \sum_{-J}^{J} (-J + J^2 x)}{\sum_{-J}^{J} (1 - Jx)}.$$

The summation over J is zero because the values run from $+J$ to $-J$, and since the sum of the squares of the first J natural numbers is $\frac{1}{6}J(J+1)(2J+1)$,

$$m_v = \frac{N_v g \beta x J(J+1)(2J+1)}{3(2J+1)},$$

and so

$$m_v = \frac{N_v g^2 \beta^2 B J(J+1)}{3kT}.$$

The susceptibility χ_v per unit volume will therefore be given by m_v/H, i.e. $\mu_0 m_v/B$. Hence

$$\chi_v = \frac{\mu_0 N_v g^2 \beta^2 J(J+1)}{3kT}. \quad (11.9)$$

We therefore see that the magnetic susceptibility should be inversely proportional to the temperature. This relationship is known as Curie's law, and is often quoted in the form

$$\chi_v = C/T, \quad (11.10)$$

where C is called the Curie constant.

For many paramagnetic crystals this expression holds quite well (Fig. 11.3). Deviations do occur, however, and these are due to the fact that the energy level system and states of an ion in a crystal are not necessarily the same as those for a free ion. The main reason for this is that in a crystal a paramagnetic ion is surrounded by other ions with which it will interact. In particular, the neighbouring ions whether magnetic or not, will still be electrically charged and these create an electric field in the neighbourhood of the paramagnetic ion which we are considering. This is often called the crystal field and it will modify the orbits (or states) of the electrons around the paramagnetic ion so that their magnetic energy is not given exactly by (11.7). The details of the way in which the energy of these states is modified may be determined by paramagnetic-resonance experiments, as well as by more conventional spectroscopic measurements.

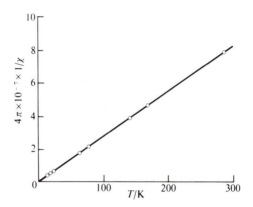

FIG. 11.3. The inverse susceptibility per kg of chromium potassium alum as a function of temperature, showing the very good agreement with Curie's law. (After W. J. de Haas and C. J. Gorter (1930). *Leiden Commun.* 208c.)

11.10. Quenching of the orbital angular momentum

Eqn. (11.9) is found to hold quite well for materials containing ions of the rare-earth group and the theoretical value of the constant C in (11.10) is in good agreement with that derived from experimental data. For ions of the iron transition group, however, the comparison is far from satisfactory. The reason for this is that for the iron group the ground state of the ions in a crystal is very different from that for the free ion. This is because the partly filled electron shell, which is responsible for producing the magnetic moment, is the outermost (3d) shell of these ions. It is therefore very strongly influenced by the crystal field which was discussed in the preceding section. The interaction is so strong that the orbital angular momentum of the electrons in the 3d shell is very often reduced to zero. This does not mean that the electrons no longer encircle the nucleus, but rather that they will be travelling around it as often in one sense as in the other sense. This effect is usually termed 'quenching' of the orbital angular momentum and it means that when the ion is in the crystal there is no orbital magnetic moment. However the crystal field hardly interacts with the spin and so the spin magnetic moment still remains.

Therefore when we calculate the susceptibility using (11.9) we should replace the total angular momentum quantum number J by that for the spin component S. When this is done, quite good agreement is obtained between the theoretical and experimental values of the Curie constant. Since L is now effectively zero, g must be set equal to 2 (eqn (11.4)). Thus for Ni^{2+} where $S = 1$, $L = 3$, and $J = 4$ (eqns (11.9) and (11.10)) yield† $C = 31.3$, but if J is replaced by S it becomes 8. This is to be compared with an experimental value of 9.7.

In the rare-earth group of ions the partially filled shell (4f) of electrons is *not* the outer shell and so it is partly screened from the crystal field by the 5s and the 5p subshells. Whilst some interaction does occur, the orbital angular momentum is not quenched and the quantum number J can be used in (11.9). This interaction, however, does modify the energy states and it often leads to a relation for the susceptibility which may be expressed as a power series

$$\chi_v = C/T + C'/T^2 + \cdots.$$

11.11. Magnetic resonance

One of the most important techniques for investigating the energy states of magnetic ions in crystals is that of magnetic resonance. We give here a very brief outline of the fundamentals which are involved in these measurements, but the reader is referred to more specialized texts for details.

The basic principles can be clearly explained by considering the case of the simplest type of paramagnetic ion—one which, in the absence of a magnetic field, has a degenerate doublet ground state with no orbital moment and with a spin $S = \frac{1}{2}$. When a magnetic field B is applied this degeneracy is removed and the state splits into two levels (Fig. 11.4(a)) which are separated from one another by an energy $g\beta B$ (11.7). If at a particular value of B the sample is subjected to a radio-frequency field at an angular frequency ω such that the quantum of energy $\hbar\omega$ is just equal to $g\beta B$ then it can induce transitions between one state and the other which can result in a net absorption of r.f. energy. This absorption can be detected in the driving circuit. Clearly since

† In units of $\mu_0 N_v \beta^2 / 3k$.

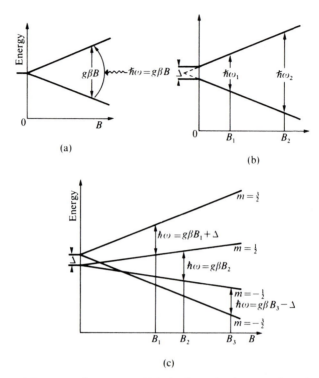

FIG. 11.4. Paramagnetic resonance. (a) A simple two-level energy scheme for a para-
magnet in a magnetic field B. Resonance occurs when the energy difference between
the levels is equal to $\hbar\omega$. (b) If the energy levels are split by Δ in zero field, then Δ
may be determined by finding the resonance fields, B_1, B_2, for two different frequencies
ω_1, ω_2. (c) Two doublets which have a separation of Δ in zero field will for a given
ω display resonances at three fields. Note that the selection rule for a transition is
$\delta m = \pm 1$ and this prevents other transitions, which are energetically possible, from
being observed.

$\hbar\omega = g\beta B$ a knowledge of B and ω enables the effective value of g to be calculated. An important interest is to study the value of g along the various
crystal axes since it can be very anisotropic. This anisotropy gives information
about the internal crystal field.

If the measurements are taken at two or more different frequencies it is
simple to decide whether or not there is any zero-field splitting of the ground
state levels (Fig. 11.4(b)), which in certain circumstances can be produced
by the internal crystal field. If there are more than two levels in the system
several absorption (or 'resonance') lines will be observed (Fig. 11.4(c)) which
can be used to deduce further details of the ionic states.

In solid state physics the most common type of resonance experiment uses a radio-frequency field at microwave frequencies ($\sim 10^{10}$ Hz) to investigate the electron spin states of paramagnetic ions. This is called paramagnetic resonance or electron spin resonance (e.s.r.). It is also possible to induce transitions with a high frequency *electric* field in the form of ultrasonic waves in the crystal (acoustic paramagnetic resonance, or a.p.r.).

Apart from paramagnetic ions e.s.r. can be used to investigate any other systems containing unpaired electrons. These occur, for example, in various defects in crystals, such as the colour centres discussed in section 3.4 (p. 39), and also in defects produced by irradiation (X-rays, gamma-rays, electron or neutron bombardment). A similar principle can be used to investigate the states associated with the *nucleus* of an atom which if it has a spin, and hence a magnetic moment, can also be made to undergo transitions from one state to another in an appropriate magnetic field. This is called nuclear magnetic resonance (n.m.r.). Because the nuclear magnetic moment is so much smaller (about 2000 times) than the electron moment, n.m.r. utilizes much lower frequencies—around 10^7 Hz.

Resonance techniques have also been devised to investigate spin wave energies (section 12.6, p.203) and the effective mass of electrons both in semiconductors (cyclotron resonance) and in metals (Azbel–Kaner resonance).

11.12. The conditions for observing resonance

In order for resonance to occur the quantum $\hbar\omega$ must clearly 'match' the energy separation of the particular levels which are being investigated, but this is not the only criterion which must be satisfied. Quantum mechanics shows that in the presence of a radiation field, transitions not only occur from the lower to the higher state (with the absorption of energy), but transitions downwards from the upper to the lower state, with the *emission* of energy, are equally probable ('induced emission'). In order, therefore, that there should be a *net absorption* of energy it is necessary to ensure that there are always more ions in the lower state than in the upper one (Fig. 11.5(a)). This is usually achieved by cooling the specimen to liquid hydrogen or helium temperatures. However, the absorption of radiation itself will tend to increase the population of the upper state thereby cutting down the absorption signal. There must also be, therefore, a mechanism for continually reducing this population by interaction with the lattice phonons so that rapid relaxation to the ground state can occur with the emission of a phonon (Fig. 11.5(b)).

Finally, even if all these conditions can be satisfied, the transition will only occur if the 'selection rules' of quantum mechanics are satisfied. These can be complicated, although for the standard paramagnetic resonance experiment there is one basic rule—the magnetic quantum number m can only change by ± 1. This rule considerably limits the number of lines which would otherwise be observed from a complicated level scheme.

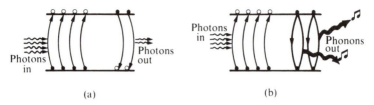

FIG. 11.5. Conditions for observing resonance. In a given field electrons are excited from the lower to the upper level by absorbing photons from the r.f. circuit. They can drop down to the lower level again by emitting another photon (a), in which case the overall absorption is reduced. Ultimately the population of upper and lower levels will become equal and there will be no net absorption. If, however, a phonon is emitted (b), in preference to a photon, the absorption signal is unimpaired.

11.13. Saturation

If an excited particle is completely isolated it cannot emit a *phonon*. This is obviously true for an ion in a vacuum, but it also applies in a solid, to a certain extent at least, if the interaction with the phonons is very weak. If this situation exists in a resonance experiment, then not only does the excitation to the higher state occur by photon absorption, but the relaxation to the ground state also occurs, either completely or preferentially, by photon emission, as in Fig. 11.5(a). Since the probabilities for photon-induced emission and absorption are equal, the ultimate result is that the populations of the upper and the lower levels of the system will equalize and so the resonance signal will disappear. This is the phenomenon of *saturation*. It is readily observable at room temperature in nuclear magnetic resonance, since the nucleus, being well within the atom, interacts only very weakly with the phonon system. It may also be observed, however, in electron spin resonance, usually at low temperatures, because the number of phonons which are present is then very small.

The saturation effect is not always complete—it depends on the strength of the phonon coupling with the excited state and so it sometimes shows up as a diminution, or possibly a complete disappearance of the resonance signal after a short time, whereas in other cases the resonance signal is always present, but its strength does not increase in proportion to the input driving power.

PROBLEMS

11.1. Calculate the magnetic susceptibility at 300 K for a salt containing one g mole of Cr^{2+} ions. Cr^{2+} has four electrons in the 3d shell.

11.2. The ion Dy^{3+} has nine electrons in the 4f shell. What are the values of L, S, and J? Calculate the susceptibility at 4 K of a salt containing 1 g mole of Dy^{3+}

11.3. A magnetic field is applied to a salt containing Cu^{2+} ions. Cu^{2+} has nine electrons in the 3d shell. What magnetic field must be applied to a salt containing Cu^{2+} when at 1 K, so that 99 per cent of the ions are in the lowest energy state?

11.4. A paramagnetic ion in a crystal has a ground state $S = \frac{3}{2}$ and the orbital moment is quenched. The state is split by the crystalline electric field into two degenerate doublets, the upper with $S_z = \pm\frac{3}{2}$ and the lower with $S_z = \pm\frac{1}{2}$. These are separated in zero magnetic field by an energy Δ. A paramagnetic resonance experiment shows three resonances at fields B_1, B_2, and B_3. Draw a diagram showing how the levels are split in a magnetic field and indicate which transitions are observed. The signals at the two lower fields are found to occur at 0·045 T and 0·3 T. If the resonance frequency is 10^{10} Hz calculate the zero-field splitting Δ and the electronic g-factor for the ion. At what field should the third resonance be found?

11.5. A substance containing isolated paramagnetic ions, each having unit angular momentum, is placed in a magnetic field B at a temperature T. Write down an expression for the magnetic energy of the system and hence derive a formula for the magnetic contribution to the specific heat. Show that at low temperatures, where the magnetic energy of an ion is much greater than kT, this specific heat rises exponentially with T, whereas at high temperatures it decreases as T^{-2}.

11.6. The susceptibility of a paramagnetic salt can be used as a thermometer to measure very low temperatures. A dilute nickel salt to be used as a thermometer contains 10^{27} Ni^{2+} ions per cubic metre. A primary and a secondary coil are wound tightly around the crystal. Reversal of a current in the primary circuit induces an e.m.f. V across the secondary terminals. Estimate the ratio of the values of V produced if the same current is reversed, first at 0·1 K and then at 1 K. (Assume $g = 2$, $S = 1$, orbital moment quenched.)

12. Ferromagnetism, antiferro-magnetism, and ferrimagnetism

B Y far the oldest solid state phenomenon which has been exploited is the ability of a piece of magnetite to orient itself in the earth's magnetic field. This is an example of *ferrimagnetism* and it is one of the set of three magnetic phenomena which are produced by *interactions* between the magnetic ions in a solid.

12.1. Interactions between magnetic ions

Materials which are ferromagnetic or ferrimagnetic can possess a magnetic dipole moment even in the absence of an applied magnetic field.† This is quite unlike the behaviour of paramagnets, discussed in the previous chapter, which need to have their elementary dipoles oriented by an external field before they display any magnetic moment. The phenomena which we now describe are due to interactions between the magnetic ions‡ which are strong enough to yield a mutual alignment of magnetic moments.

Any interactions between ions which tend to produce an ordering will only become important if the strength of the interaction is large compared with those mechanisms which can disrupt the ordering. In particular, the ordering interaction must be large compared with the thermal vibrational energy which will of course tend to disorder any alignment.

The only orderings which we shall consider are those in which the dipole moments of neighbouring ions are either parallel or antiparallel to one another. The parallel arrangement occurs in ferromagnets and the antiparallel ordering is typical of antiferromagnets and ferrimagnets (ferrites) (Fig. 12.1).

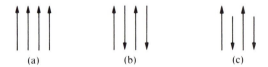

(a)   (b)   (c)

FIG. 12.1. The three simplest ordering arrangements in magnetic materials. (a) Ferromagnetic, (b) antiferromagnetic, and (c) ferrimagnetic.

† Common experience with a piece of iron would suggest that it only has a magnetic moment if it is first 'magnetized' by an external field. This, however, is a consequence of the formation of domains of opposing magnetization (section 12.7). Individual domains do possess a magnetic moment even in the absence of a field.

‡ In our discussion on paramagnetism we tacitly assumed, when using the Maxwell–Boltzmann function in (11.8) that the ions are completely independent of one another and that there are no interactions between them.

12.2. Dipole–dipole interactions

The most obvious type of interaction between magnetic ions is the effect of the magnetic field of one dipole on its neighbour. However, this is a very weak interaction which can be completely disregarded, except at very low temperatures, as the following simple calculation shows. The intensity, H, at a distance d from a dipole of strength m_m is of the order of $m_m/(4\pi d^3)$ and the magnitude of the magnetic energy due to H of a similar dipole is of the order of

$$m_m B = \mu_0 m_m H = \mu_0 m_m^2/(4\pi d^3).$$

If m_m is the Bohr magneton (0.9×10^{-23} A m^2) and d is the interatomic spacing (~ 0.3 nm), this energy is $\sim 3 \times 10^{-25}$ J. This is equivalent to the thermal energy kT at a temperature of about 10^{-2} K. Hence the mutual alignment of moments due to direct magnetic interaction can only take place at very low temperatures.

12.3. Exchange interaction

It is however well known that magnetic ordering in ferromagnets can occur well above room temperature, e.g. in Fe up to 1043 K. To account for this it has been necessary to postulate an *exchange interaction* which in essence is the same as that which is used to explain co-valent bonding (section 1.16, p.14).

Exchange interactions are a consequence of the quantum nature of the system and they have no classical counterpart. The effect can be viewed by the following argument. If the wave functions of two atoms overlap, then the electrons of atom 1 are to some extent also associated with atom 2, and vice versa. Thus there must be some interaction between the two groups of electrons because they can exchange their roles. In particular there is a correlation between their spins—and hence also their dipole moments. In most materials the system seems to have a lower energy if these spin moments are antiparallel. It is only in ferromagnetic materials that the parallel orientation has a lower energy. The magnitude of the exchange interaction depends on the amount of overlap of the electron wave functions and this makes it a very difficult problem to calculate with any confidence. In the ferromagnetic metals it is thought that interaction via the conduction electrons is also an important factor in producing the mutual alignment of spins.

The complexity of the phenomenon can be further appreciated by calculating the contribution in Bohr magnetons (β) per atom to the measured saturation magnetization, i.e. when all the spins are aligned. This is always non-integral (e.g. for Fe it is 2.2 β per atom) and it suggests that an interaction between electrons in energy bands is involved, rather than an alignment of individual spins.

12.4. Ferromagnetism; the internal (or molecular) field

Ferromagnetic materials are those which at high temperatures behave as ordinary paramagnets but which spontaneously acquire a permanent magnetic dipole moment at a critical temperature which is known as the Curie temperature T_C. The most common ferromagnets are the metals iron, cobalt, and nickel and some of their alloys.

The alignment of the elementary dipoles may be treated in terms of the exchange interaction as follows.

The energy of ion i (with spin S_i), due to exchange interactions with its neighbours $j_1, j_2, ...$ may be written in the form

$$E_{ex} = -S_i \sum_j J_{ij} S_j, \qquad (12.1)$$

where the summation is over all the j, and J_{ij} is the strength of the interaction.

If we assume (as is true for the iron group) that the magnetic moment m_i of an ion is due to spin alone, then $m_i = -g\beta S_i$, and hence substituting for the S we may write

$$\begin{aligned}
E_{ex} &= (g\beta)^{-2} m_i \sum_j J_{ij} m_j \\
&= m_i \, \overline{m} \overline{J}/(g\beta)^2,
\end{aligned} \qquad (12.2)$$

where $\overline{J}$ is an average of the J_{ij} and $\overline{m}$ is the effective vector summation of the m_j over the j neighbours of the ith ion.

This expression is of the form $m_i B_{int}$, and hence the exchange energy of ion i can be considered to be due to the magnetic energy of the dipole m_i when it is in a field B_{int} whose strength is proportional to $\overline{m}$, the average dipole moment of the material. The better the alignment of the dipoles, the greater will be this 'internal' field and hence the stronger will be its influence to produce further alignment. The onset of ferromagnetism is called a *cooperative transition* because the very change (i.e. the alignment) which occurs during the transition helps to accelerate the transition itself. Cooperative transitions are characterized by a change in properties which occur over a very narrow temperature range. They are usually accompanied by a sharp peak in the specific heat.

It must be emphasized that the internal magnetic field which we have introduced is fictitious. It is only used because the exchange interactions themselves (which are actually due to *electrostatic* forces) are outside our ordinary experience.

12.5. The behaviour of a ferromagnetic above the Curie temperature

Above the Curie temperature the material is paramagnetic, and we may obtain an expression for the temperature-dependence of its susceptibility.

If we express the proportionality of the internal field B_{int} to the magnetization m_v in the form

$$B_{int} = \lambda m_v,$$

then the total effective field acting on the material when it is in an external field of magnitude B is $B + \lambda m_v$. If we assume that Curie's law (11.10) holds, we then have

$$m_v = C(B + \lambda m_v)/(\mu_0 T), \tag{12.3}$$

and hence

$$m_v(\mu_0 T - \lambda C) = CB.$$

The susceptibility is therefore

$$\chi_v = \mu_0 C/(\mu_0 T - \lambda C)$$

or

$$\chi_v = C/(T - \theta_W), \tag{12.4}$$

where $\theta_W = \lambda C/\mu_0$. This is the Curie–Weiss law, and θ_W is called the Weiss constant. At temperatures greater than θ_W the substance has a temperature-dependent paramagnetic susceptibility, and this describes the experimental results quite well. We see that at $T = \theta_W$ (12.4) breaks down and we can associate θ_W with the Curie temperature. The agreement between θ_W and the temperature at which ferromagnetism sets in is usually quite reasonable.

12.6. Ferromagnets below the Curie temperature; spin waves

Whilst we have stated that below the Curie temperature all the magnetic dipoles are spontaneously aligned, this clearly cannot be strictly true because the thermal vibration will always tend to produce some disorientation. At very low temperatures ($T \ll T_C$) the magnetization of a ferromagnet is found to vary in the following manner

$$m_v = m_0(1 - aT^n), \tag{12.5}$$

where m_0 is the magnetization at 0 K and $n = \frac{3}{2}$ (Fig. 12.2). This temperature dependence does not follow directly from the internal-field theory.

The effect has been satisfactorily explained by assuming that any spin misalignment which arises will not be static. If at some temperature above 0 K we could look at the assembly of dipoles we would see a complicated haphazard pattern in which the spins are continually changing their orientation. This picture can be simplified by considering the changing alignments as a superposition of a set of single misorientations each of which is travelling

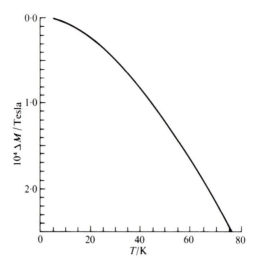

FIG. 12.2. The variation of the magnetization of nickel in a field of 0·1 T at low temperatures. (After B. E. Argyle, S. H. Charap and E. W. Pugh (1963). *Phys. Rev.* **132**, 2051.)

from one ion to the next through the lattice. The propagation of a single misorientation is called a *spin wave*. It will be seen that the idea is analogous to the model which is used to analyse the complicated thermal motion of atoms in a solid as a superposition of normal modes (section 5.7, p. 84). A full development of spin-wave theory is able to account for the $T^{\frac{3}{2}}$ behaviour of the magnetization in (12.5).

By analogy with phonons and photons, the excitations associated with spin waves are also quantized and the individual excitations or quanta are called *magnons*. They obey Bose–Einstein statistics and their contribution both to the specific heat and to the thermal conductivity has been detected at low temperatures. They may also be investigated by spin-wave resonance experiments and by inelastic neutron scattering.

12.7. Ferromagnetic domains

It is common knowledge that a ferromagnet does not always seem to possess a magnetic moment. It needs to be 'magnetized' by putting it in an external magnetic field (which may be very small compared to the 'internal field') and, depending on the material, the moment is to a greater or lesser degree retained when the field is removed. The magnetization can be destroyed if the material is subjected to mechanical shock or to heat.

This behaviour is explained by the fact that below the Curie temperature the material is subdivided into 'domains', each of which is spontaneously magnetized in accord with (12.5) but in which the direction of magnetization

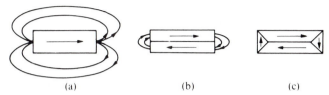

FIG. 12.3. Ferromagnetic domains. (a) If the whole specimen is magnetized in one direction there is a large amount of magnetic energy stored in the field outside the specimen. (b) A pair of domains magnetized in opposite directions reduces this external field and energy. (c) Small closure domains at the ends of the specimen further reduce the magnetic energy.

changes between one domain and the next. The explanation for the existence of these domains is clear from Fig. 12.3(a). If the material is all magnetized in the same direction (as in a bar magnet) there is a large amount of magnetic energy ($\frac{1}{2}H . B$ per unit volume) stored in the free space around the magnet. If the material is split into two domains of opposite magnetization (Fig. 12.3(b)) this energy is greatly reduced. Further domains (Fig. 12.3(c)) are possible which reduce the field emerging from the ends of the specimen. These are called closure domains. The material does not split up into an infinitely fine mosaic of domains because the formation of domain boundaries (sometimes called Bloch walls), where the orientation of the moments changes fairly rapidly, requires extra energy. The process of magnetization of a ferromagnet entails a change in the arrangement of the domains so that their magnetization becomes more nearly parallel.

The presence of domains can be demonstrated and studied experimentally by covering the surface of the ferromagnet with very fine iron powder. This delineates the boundaries of the domains and they can be observed under the microscope (Fig. 12.4). The presence of domains also introduces two other mechanisms whereby the energy of the system is increased and which therefore tend to limit the domain system. These are dealt with in the next two sections.

12.8. Crystalline anisotropy energy

It is found from experiments on single crystals that it is easier to magnetize them along one axis than along another, e.g. for Fe the [100] axis requires less field than the [110] and this in turn is easier than the [111]. For Ni the easy axis is [111] whereas for Co it is the hexagonal axis. The anisotropy energy† is the extra energy per unit volume which is needed to magnetize the material along the 'hard' direction. The domains themselves will tend to be magnetized along the 'easy' direction but in the walls between domains some magnetization will need to be in the 'hard' direction and this will increase the energy associated with the domain boundaries (see Problem 12.3).

---

† Crystalline anisotropy should not be confused with *shape* anisotropy which depends on the shape of the specimen.

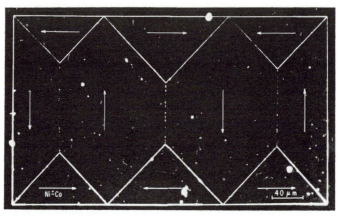

FIG. 12.4. Ferromagnetic domains in a nickel–cobalt platelet. The triangular closure domains are clearly shown. The arrows indicate the direction of magnetization. The overall length of the specimen is about 0·5 mm. (Photograph by courtesy of R. W. DeBlois.)

12.9. Magneto-striction

When a material is magnetized its dimensions change slightly (the change may be either positive or negative). This is called magneto-striction. In any arrangement of domains the dimensional changes between neighbours will not match and so the magneto-striction gives rise to elastic strain energy which will also limit the production of domains.

Very small particles whose dimensions are less than the width of a domain wall will tend to remain as single domains because the energy to form extra domains will be greater than the field energy around the particle.

12.10. The hysteresis curve

When a ferromagnet is placed in a magnetic field which is cycled from zero up to a maximum positive value, back to zero, and then to a negative value, and so on, the magnetization and the value of B within the material follow a typical hysteresis curve as shown in Fig. 12.5(a). If the maximum intensity is sufficiently large the material will attain the saturation magnetization given by (12.5). After a cycle is completed and H is again zero the material still retains some magnetization and this is called the remanence (i.e. the intercept on the positive B axis). In order to reduce B to zero, a reverse intensity, the coercive force (i.e. the intercept on the negative H axis) is required. The energy dissipated during the cycle is equal to $\oint B \, . \, dH$, the area enclosed by the curve.

What processes occur during the hysteresis cycle? The main mechanisms seem to be as follows (Fig. 12.5(b)): (1) On starting from zero the applied

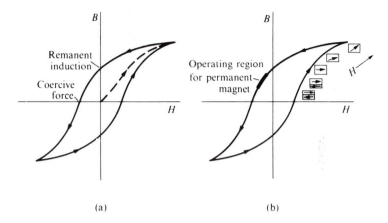

(a)   (b)

FIG. 12.5. The hysteresis curve for a ferromagnet. (a) The general form showing the remanent induction and the coercive force. (b) The domain motion that occurs during the positive portion of the curve. Note also that due to demagnetization effects the operating point for a permanent magnet is in the upper negative quadrant of the curve (see text).

field moves the Bloch walls between domains in such a direction that the domains which are more nearly magnetized in the field direction grow at the expense of those which are not so favourably oriented. For small fields and displacements this wall motion is *reversible*. (2) As H is increased further, the Bloch walls have to be pulled over various obstacles—impurities, crystal defects, dislocations, and so on. If this involves surmounting a potential energy maximum (Fig. 12.6) the wall cannot easily return to its original position when H is reduced, and so this is an *irreversible* process. (3) Eventually all

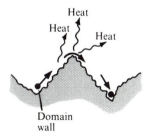

FIG. 12.6. Diagram of a domain wall moving over a potential barrier. Energy is expended irreversibly as heat.

the domains which were magnetized in the 'wrong' direction will be eliminated. A further increase in H will tend to rotate the direction of magnetization of the domains away from their 'easy' orientation so that they are more nearly parallel to the field direction.

These three stages are of course not sharply defined. There will be an overlap of the mechanisms at various parts of the cycle.

12.11. Demagnetization

It should be noted that it is only when a field is applied parallel to the axis of a long thin specimen that the external field is equal to the actual field acting on the domains within the material and hence care has to be taken in measuring and interpreting a B versus H curve. In particular a permanent magnet in zero external magnetic field has a value of H within the material which is in the opposite direction† to B.

The operating point of such a magnet is therefore somewhere within the negative upper quadrant of the hysteresis curve (Fig. 12.5(b)).

12.12. Hysteresis curves for various applications

Hysteresis curves with different characteristics are necessary for various applications. Permanent magnets require a large remanence and coercive force so that they retain their magnetization even when maltreated (Fig. 12.7(a)). They are made from fine-grained alloys containing non-magnetic inclusions that tend to inhibit domain wall movement. The magnetic

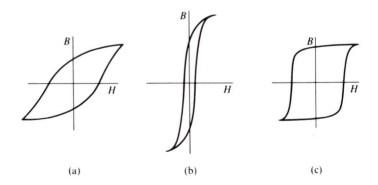

(a)   (b)   (c)

FIG. 12.7. Types of hysteresis curve for various applications. (a) Permanent magnet. (b) Transformer lamination. (c) Ferrite store.

† This can be deduced by recalling that the line integral $\oint H \cdot dl = 0$, when taken around a closed loop. Since B is continuous and H must be parallel to B in the air gap, it follows that H is antiparallel in the material.

particles are often rod-shaped and are aligned along the direction of magnetization; this hinders demagnetization. Alnico 5, a high grade magnet alloy (Fe 51, Co 24, Ni 14, plus Al, Cu, Nb) is of this type. It has a coercive force of 5×10^4 A m^{-1} compared with ~ 50 A m^{-1} for Fe, although its remanence, at ~ 1.2 T, is almost the same. However the most outstanding material is SmCo$_5$, with a coercive force of $\sim 6 \times 10^5$ A m^{-1} and a remanence of ~ 0.9 T.

The magnetic cores of transformers and electric motors require quite different characteristics. They need very thin, high, hysteresis curves (Fig. 12.7(b)) to avoid saturation, to minimize losses, and to ensure that there is maximum flux linkage between primary and secondary circuits. In order to reduce eddy current losses the cores incorporate impurities to increase their electrical resistance and they are also laminated. They should have large oriented crystallites so that the domains are large and their walls move easily. Grain oriented FeSi 4 per cent is a popular material for power transformers. The Si increases the resistivity and it also stabilizes the b.c.c. phase of Fe during manufacture so as to ensure the formation of large crystallites. For higher frequencies alloys containing Ni (Permalloy, Supermalloy) are used.

Magnetic screening also requires very 'soft' materials with a very high permeability in which there is little magnetic anisotropy or magnetostriction, so that the domain walls can move and respond rapidly to outside fields. Alloys of the rough composition FeNi 75 per cent (mumetal, Permalloy, Supermalloy) have permeabilities between 10^5 and 10^6 and are used for this purpose.

Early computers had magnetic memory stores. These were made of ferrite (section 12.16) and they required a well-defined switching field with a very sharp change between the positive and negative magnetizations. Hence they needed a very square hysteresis curve (Fig. 12.7(c)).

For recording tape the most desirable characteristics are that the magnetization should be retained for a long time and that it should be as nearly as possible proportional to the applied field. To minimize demagnetization the magnetic particles on the tape are therefore usually needle-shaped, lying parallel to the recording direction.

12.13. Antiferromagnetism

The exchange energy is very sensitive to the spacing of the paramagnetic ions and it appears that rather special conditions are required for the energy to be a minimum when the neighbouring spins are parallel. In most cases the exchange energy is minimized when neighbouring spins are antiparallel. The alignment of spins in an antiparallel array is also a cooperative transition and it occurs at a temperature known as the Néel temperature, T_N. This is the phenomenon of antiferromagnetism. The ordered moments can be considered to lie on two similar interpenetrating sub-lattices (Fig. 12.8) within

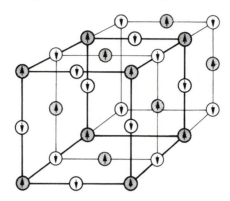

FIG. 12.8. A typical antiferromagnetic structure showing two interpenetrating sub-lattices which are magnetized in opposite directions.

each of which the moments are all parallel, with the spins on one sub-lattice being antiparallel to those on the other.

Since there is no net magnetization at the onset of antiferromagnetism, the transition is not so dramatic as in ferromagnets. However, neutron diffraction (section 2.18, p. 34) clearly shows the strengthening of some peaks and the development of others at T_N due to magnetic ordering.

If the two sub-lattices are magnetically similar to one another the total magnetic moment will be zero. If, however, the two sub-lattices are magnetically inequivalent then the transition results in a material which does have a permanent magnetic moment and these substances are called *ferrimagnets* or *ferrites*.

The only well known metals which are antiferromagnetic are Cr ($T_N = 475$ K) and α-Mn ($T_N = 100$ K). The lighter rare earth metals from Ce to Eu become antiferromagnetic at low temperatures (T_N between 10 and 87 K); Gd is ferromagnetic, but the heavier rare earth metals from Tb to Tm show complex behaviour. They are ferromagnetic at high temperatures and they become antiferromagnetic at lower temperatures. The list of compounds which become antiferromagnetic, however, is very large as most paramagnetic substances become antiferromagnetic at a sufficiently low temperature.

12.14. The behaviour of an antiferromagnet above T_N

The paramagnetic properties above T_N can be treated by an internal-field model, akin to the one which we used for ferromagnetism. We now assume that an ion on sub-lattice X interacts with its neighbours (on Y) by means of an internal field B_{iY}. If m_Y is the volume magnetization of lattice Y, and B is the external field, then the effective field B_X on X is

$$B_X = B + B_{iY} = B - \lambda m_Y$$

and similarly

$$B_Y = B + B_{iX} = B - \lambda m_X.$$

The negative signs are used because the exchange interaction will tend to destroy the alignment to B. Above T_N we can calculate the magnetization for each lattice separately, using B_X and B_Y instead of $B + \lambda m_v$ in (12.3). Thus

$$m_X = \tfrac{1}{2}C(B - \lambda m_Y)/(\mu_0 T) \text{ and } m_Y = \tfrac{1}{2}C(B - \lambda m_X)/(\mu_0 T).$$

Then

$$m_v = m_X + m_Y = \tfrac{1}{2}C\{2B - \lambda(m_X + m_Y)\}/\mu_0 T.$$

Hence

$$\chi_v = \mu_0 m_v/B = C\{1 - \lambda(m_X + m_Y)/2B\}/T \approx C/(T + \theta'), \tag{12.6}$$

where

$$\theta' = \lambda C/(2\mu_0). \tag{12.7}$$

Thus above the Néel temperature an antiferromagnet obeys a Curie–Weiss law (12.4) but with a change of sign in the denominator.

It can be shown that θ' can be interpreted as being the temperature at which the cooperative transition occurs, although θ' deduced from measurements in the paramagnetic region usually turns out to be two or three times higher than the experimental value of the Néel temperature.

12.15. The susceptibility of an antiferromagnet below T_N

In the previous section we have shown that the susceptibility of an anti-ferromagnet above T_N follows a Curie–Weiss law (12.6). In a cubic crystal this should be isotropic but below T_N there is strong anisotropy. If for example, we assume (Fig. 12.9) that an external field B is applied at right angles to the

Fig. 12.9. The susceptibility of an antiferromagnet below the Néel temperature when the field is perpendicular to the alignment axis (see text).

direction of magnetic alignment the field will tend to rotate the magnetization on both sub-lattices towards B by the same amount, and it can be shown that the susceptibility in this direction, $\chi_\perp$, is independent of T and has a value of about μ_0/λ (see Problem 12.6).

However, if B is applied in a direction parallel to the spins a very different situation develops. At $T = 0$ K the spins on the two sub-lattices are in perfect alignment and hence the application of a parallel field will produce no resultant moment on them and $\chi_\parallel$ will be zero. As the temperature is raised the alignment is upset slightly and so B will be able to produce a small rotation of the spins and hence a small susceptibility. As T_N is approached, $\chi_\parallel$ becomes larger, tending towards μ_0/λ (Fig. 12.10).

A powdered specimen of an antiferromagnet will have a susceptibility below T_N which will be an average of $\chi_\parallel$ and $\chi_\perp$, given by

$$\chi_{\text{powder}} = (\chi_\parallel + 2\chi_\perp)/3. \tag{12.8}$$

Thus at 0 K χ_{powder} will be equal to two-thirds of its value at T_N.

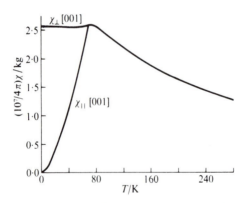

FIG. 12.10. The susceptibility of MnF_2 parallel and perpendicular to the [001] axis of the crystal. This shows the very large anisotropy of an antiferromagnet below the Néel temperature. (Data by courtesy of Dr. S. Foner.)

12.16. Ferrimagnetism

In some crystals with a more complicated structure, the magnitude of the magnetic moments on each of the two sub-lattices are not exactly the same. Thus when spontaneous antiparallel alignment occurs, the material, instead of having zero magnetic moment, has a net permanent magnetization. This phenomenon is known as *ferrimagnetism*. The most common material which shows this behaviour is magnetite, Fe_3O_4, which has a spinel structure. Materials which exhibit ferrimagnetism now all tend to be called *ferrites* even if they do not contain any iron.

Ferrites are of great technical importance because they have a very high electrical resistivity together with a high magnetic permeability. They can therefore be used as cores of inductive elements which operate at frequencies which are very much higher than those for which iron cores are employed. Their hysteresis curves made them very suitable for memory stores (section 12.12) and they are also used as directional elements in microwave equipment.

When they are magnetized their behaviour is very similar to that of ferromagnets. They have a domain structure and their properties can be treated using conventional spin-wave theory.

Their saturation magnetization behaviour as a function of temperature is complicated because the magnetic moments of the two inequivalent sublattices usually do not each have the same temperature-dependence.

PROBLEMS

12.1. Magnetic ions with $S = \frac{5}{2}$ and $L = 0$ are spaced 0·5 nm apart. Calculate the magnetic energy of one ion due to the field of its neighbour. At approximately what temperature would the moments be aligned due to this type of interaction?

12.2. Derive the relationship between the 'internal' field of a ferromagnet and the Curie temperature. Calculate this field for iron which has a Curie temperature of 1043 K and an effective moment of 2·2 Bohr magnetons per ion.

12.3. Calculate the width of a domain wall as follows. Across a domain wall the direction of magnetization changes by 180°. If the wall is N atoms thick write down an expression for (a) the exchange energy per unit area of wall and (b) the anisotropy energy assuming that half the wall is magnetized in the hard direction. Minimize the total energy with respect to N. If the exchange constant $J = 10^{-21}$ J, the anisotropy energy $K = 4 \times 10^4$ J m^{-3} and the atomic spacing is 0.3 nm, calculate the width of the wall and the energy per m^2 of wall for $S = 1$.

12.4 The ferromagnet europium oxide has a Curie temperature of 70 K. The europium ion has $J = \frac{7}{2}$ and $g = 2$. Assuming the internal-field model, determine the ratio of the magnetization at 300 K in a field of 10^{-2} T to that at 0 K.

12.5. The hysteresis curve in Fig. 12.7(b) has a width H of 50 A m^{-1} and an overall length in B of 1 T. Estimate the energy per cycle which is dissipated per cubic metre. If the material is used as a cubic transformer core of side 10 cm, estimate the power dissipation at 50 Hz.

12.6. Show that the susceptibility of an antiferromagnet below the Néel temperature is μ_0/λ when the field is applied perpendicularly to the axis of alignment. This may be done as follows. In Fig. 12.9 draw the direction of the internal field produced by m_X (remember that it acts in the direction opposite to m_X). This field produces a moment on m_Y which acts in opposition to that due to the external field B. Calculate the angle for which the net moment is zero. For this angle determine the component for m_Y (and m_X) along B and hence find the susceptibility. Assume that rotations from the axis of antiferromagnetic alignment are small.

13. Dielectric properties

13.1. Introduction

THE effect of an electric field on the properties of solids is not usually given so much emphasis as the magnetic phenomena which we discussed in previous chapters—and yet it is arguable that dielectric effects are in many respects of more practical importance. In particular we should mention the propagation through dielectrics of electromagnetic radiation (from radio-frequencies to beyond the visible range), the dispersion of light in optical media, dielectric losses and breakdown in capacitors, and the widespread use of piezoelectric devices.

13.2. Polarization

When a system is subjected to an electric field, $\mathscr{E}$, there is a tendency for the positive and negative charges (i.e. the nuclei, the electrons, or the ions) within the material to be displaced relatively to one another so that the system acquires an electric dipole moment. The dipole moment per unit volume is called the polarization, P.

It is an experimental fact that this moment is almost exactly proportional to the field for all normal values of $\mathscr{E}$, although the fields which are produced by high-power lasers are large enough to give rise to non-linear behaviour. If we neglect nonlinearity we may define an electric susceptibility† χ_e by the relation

$$P = \varepsilon_0 \chi_e \mathscr{E}, \tag{13.1}$$

where ε_0 is the permittivity of free space.

The basic experiment on a dielectric is the measurement of the dielectric constant ε which is defined as $(1 + \chi_e)$ (see section 13.7). This is usually done by measuring the capacitance of a condenser with and without the dielectric between its plates. The ratio of these two capacitances is ε (section 13.8).

The theory of dielectrics is not in a very satisfactory state and there is very little of practical value that can be derived from first principles. For this reason when we describe the various mechanisms for polarization we shall not go into any mathematical detail. Also most of the mechanisms which have been proposed have been verified on gases and liquids (these latter provide the most dramatic effects) and it is not entirely clear whether some of these are operative in solids.

---

† There are, of course, very strong analogies between various parts of dielectric theory and the corresponding effects and theorems in magnetism, but we leave it to the reader to make the necessary associations.

13.3. Polarization mechanisms

We now describe briefly the three different mechanisms which can result in the polarization of a medium.

(1) The simplest effect occurs when an electric field acts on an individual atom. In zero field the positively charged nucleus and the negative cloud of electrons are symmetrically disposed to one another and the atom has no dipole moment. When a field $\mathscr{E}$ is applied the nucleus and the electrons move with respect to one another (Fig. 13.1) and the atom acquires a dipole moment p, which we shall assume to be proportional to $\mathscr{E}$, so that we may write

$$p = \alpha \mathscr{E} \dagger \tag{13.2}$$

where α is the polarizability of the atom.

This is called *electronic* or *induced* polarization and it is operative in all dielectric materials.

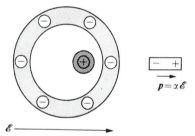

$$p = \alpha \mathscr{E}$$

Fig. 13.1. Polarization of an atom in an electric field. The electron cloud and the nucleus move with respect to one another so that the atom has a dipole moment.

(2) If the system is composed of heteronuclear molecules then the disposition of the individual atoms within a molecule may be such that the molecule itself has a *permanent* dipole moment. These are called polar molecules. A molecule which is composed of different atoms is not necessarily polar, e.g. CO_2 is non-polar because the carbon and oxygen atoms are arranged in a straight line with the carbon in the middle (Fig. 13.2(a)). H_2O is polar because the ions are arranged in a triangle (Fig. 13.2(b)).

In zero field the permanent dipoles will be randomly oriented and the system has no net polarization, but an electric field will tend to align the dipoles and the material will acquire a net moment. This is called *orientational polarization*. Due to the randomizing effect of the thermal vibrations this type of polarization is more effective as the temperature is decreased and it gives rise to a

$\dagger$ p is sometimes defined as $\alpha \varepsilon \mathscr{E}$.

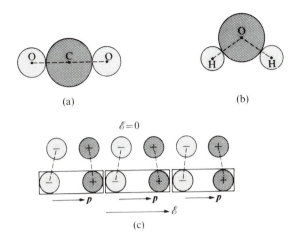

FIG. 13.2. CO_2 does *not* have a permanent dipole moment (a) because the atoms are in line, whereas water (b) does have a permanent moment. (c) Ionic polarization can occur in a solid when the field displaces the negative and positive ions with respect to one another. It also produces a polarization charge on the surface.

dielectric constant which is temperature dependent.† In solids, however, the molecules are usually so tightly bound that orientational polarization does not seem to occur. It is much more important in liquids and gases.

(3)   A second effect can arise with permanent dipoles. The field will tend to stretch the bonds between the ions and this will change the moment of the molecule. This is called *molecular* or *ionic* polarization.‡ A similar effect can occur in an ionic crystal such as Na^+Cl^-; an electric field can shift the positive ion sub-lattice relative to the negative ion sub-lattice and this gives the crystal a dipole moment (Fig. 13.2(c)). This is also called *ionic polarization*.

All these mechanisms yield a polarization which is proportional to the field acting on the ions, as in (13.2). The main problems in the theory of dielectrics are (a) to obtain a relation between the dielectric constant (which is a macroscopic quantity) and the microscopic dipole moment of an ion or a molecule, and (b) to be able to make a theoretical calculation of the polarizability α for particular atoms. The first task can be achieved fairly reasonably,

---

† The mathematical theory is exactly the same as that for Curie's law for paramagnetic susceptibility (11.10), and it leads to an electric susceptibility which varies as $1/(T+\text{constant})$.

‡ We have discussed orientational and ionic polarization as if they are quite separate effects. A complete treatment however should take account of the influence of one on the other.

but the second, apart from one or two special cases, such as the helium atom, has not yet been satisfactorily solved.† We are therefore not yet in a position to be able to calculate the dielectric constant of a material from first principles.

13.4. Electrostatics

In order to show the relationship between the dielectric constant and the polarizability we first need to give a brief outline of some basic electrostatic theory. Further details and proofs of the theorems will be found in any standard text.

It must first be emphasized that classical electrostatics deals with continuous media and this must always be borne in mind when we try and relate the results to effects involving individual atoms.

It is also important to appreciate the different situations which arise depending on whether the field is applied *externally* to the dielectric (e.g. as when the dielectric is between the plates of a charged condenser) and those in which the effects are initiated *within* the dielectric.

13.5. The dielectric in an external field

Let us first consider what happens when a dielectric is placed in a homogeneous electric field $\mathscr{E}_{ext}$ produced, for example, by charges placed at a great distance from the dielectric. Two effects will be operative, each of which will tend to modify the field within the dielectric.

(1) The polarization will result in net effective charges being induced on the surfaces of the dielectric (Fig. 13.3). These produce a field $\mathscr{E}_{dp}$, within the dielectric (the depolarizing field), which *opposes* the external field. The value of this field depends on the geometry of the system. We write $\mathscr{E}_{dp} = -.\mathscr{N}P$,

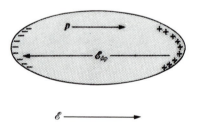

Fig. 13.3. A dielectric in an external field has a depolarizing field $\mathscr{E}_{dp}$ inside, which is produced by the surface charges on the dielectric.

† If the atom or molecule is assumed to be a conducting sphere of radius r then it is simple to show that the polarizability is equal to $4\pi\varepsilon_0 r^3$ (e.g. see B. I. Bleaney and B. Bleaney (1976) *Electricity and magnetism* (3rd edn), pp. 43–4. Clarendon Press, Oxford.

where $\mathcal{N}$ is the depolarizing factor. For a parallel-sided slab, $\mathcal{N} = 1/\varepsilon_0$; for a long thin rod with its axis parallel to $\mathscr{E}_{\text{ext}}$, $\mathcal{N} = 0$. Although important in many contexts we shall tend to disregard the effects of $\mathscr{E}_{\text{dp}}$.

(2) If the material is anisotropic, or if the field varies within the specimen, then the relative displacement of the positive and negative charges will change from one region of the material to another. This gives rise to a net effective induced charge per unit volume and it can be shown to have a value of $-\operatorname{div} \boldsymbol{P}$. It should be noted that the surface charge in (1), where $\boldsymbol{P}$ changes abruptly from zero to $\boldsymbol{P}$ at the surface, is just a special case of this induced charge due to polarization.

13.6. Free charges within the dielectric

Similar polarization effects will be produced if, instead of putting the material in an external field, we produce a field in the dielectric by introducing a free charge q within the material.

What will be the value of this field? In free space we have the result of Gauss' law

$$\int \varepsilon_0 \mathscr{E} \cdot \mathrm{d}\boldsymbol{S} = \int \varepsilon_0 \operatorname{div} \mathscr{E} \, \mathrm{d}v = q = \int \rho \, \mathrm{d}v \qquad (13.3)$$

where ρ is the free charge density.

Now Gauss' law is always valid provided we include *all* the charges within the volume of integration in (13.3). In the dielectric, as we saw from (2) above, we have an extra charge density, $-\operatorname{div} \boldsymbol{P}$. This must therefore be added to the right hand side of (13.3) to give

$$\int \varepsilon_0 \operatorname{div} \mathscr{E} \, \mathrm{d}v = \int (\rho - \operatorname{div} \boldsymbol{P}) \, \mathrm{d}v$$

or

$$\int (\varepsilon_0 \operatorname{div} \mathscr{E} + \operatorname{div} \boldsymbol{P}) \, \mathrm{d}v = \int \rho \, \mathrm{d}v. \qquad (13.4)$$

We may write therefore

$$\int \operatorname{div} \boldsymbol{D} \, \mathrm{d}v = \int \rho \, \mathrm{d}v, \quad \text{or} \quad \int \boldsymbol{D} \cdot \mathrm{d}\boldsymbol{S} = \int \rho \, \mathrm{d}v, \qquad (13.5)$$

where $\boldsymbol{D}$, the *electric displacement vector* is defined by

$$\boldsymbol{D} = \varepsilon_0 \mathscr{E} + \boldsymbol{P}. \qquad (13.6)$$

It is important to note that $\mathscr{E}$ is the effective *average* field within the dielectric when the effects of the induced polarization charges are taken into account. In uniform fields these will be, in particular, the surface depolarization charges.

Since it is based on a continuum theory this value of $\mathscr{E}$ does *not* reflect the detailed internal fluctuations of the field which must occur because of the atomic structure of the material. In general, $\mathscr{E}$ is *not* equal to the external field, although in the important case of a dielectric between the plates of a condenser with a constant potential between the plates, $\mathscr{E}$ *is* equal to the field between the plates before the dielectric was inserted. This is because the effect of the polarization charges on the surface of the dielectric is neutralized by the extra charges which they induce on the condenser plates (Fig. 13.5(b)).

The vector D is useful because Gauss' law is used most conveniently to derive D and not $\mathscr{E}$; but once D has been found it is usually necessary and very simple to calculate $\mathscr{E}$, as is shown below.

13.7. The dielectric constant

The dielectric constant ε may be defined from (13.6) and (13.1) by writing

$$(D/\varepsilon_0\mathscr{E}) = 1 + P/(\varepsilon_0\mathscr{E}) \tag{13.7}$$

$$= 1 + \chi_e \equiv \varepsilon, \tag{13.8}$$

and hence

$$D = \varepsilon\varepsilon_0\mathscr{E}. \tag{13.9}$$

The value of the dielectric constant is determined by experiment and hence $\mathscr{E}$ may always be calculated once D is known.

13.8. The parallel-plate condenser

As a simple example we calculate the capacity of a parallel-plate condenser in which a dielectric fills the space between the plates. Let the charge on each plate be $\pm q$ when a potential V is applied between them. If we draw a surface which passes through the dielectric and encloses one plate (Fig. 13.4) then we may apply the surface form of Gauss' law (13.5).

$$\int D \cdot dS = q.$$

With this simple geometry it is clear that D is constant over the surface of the plate and is normal to it, and so we may write, using (13.9)

$$AD = A\varepsilon\varepsilon_0\mathscr{E} = q,$$

where A is the area of the plate. The field between the plates is

$$\mathscr{E} = q/(A\varepsilon\varepsilon_0),$$

but the definition of capacity is $q/V = q/(\mathscr{E}d)$, where d is the distance between the plates, and hence

$$C = A\varepsilon\varepsilon_0/d. \tag{13.10}$$

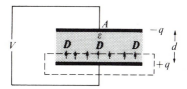

FIG. 13.4. The parallel-plate condenser (see text).

The presence of the dielectric has therefore increased the capacity by a factor ε. The comparison of the value of a capacity with and without a dielectric between its plates is the most straightforward experimental method for determining ε. For most solids ε has values between 2·5 and 15, but values as high as 10 000 may occur.

13.9. The microscopic theory of the dielectric constant

We now need to show how an external field $\mathscr{E}$ will affect the polarization of an atom within a solid.

Let us assume for the moment that we are dealing with a unit volume containing N atoms or ions which only exhibit induced polarization. Each atom has a dipole moment p given by (13.2). We wish to find the moment P of the whole assembly. Clearly this must be equal to Np; what could be simpler? The snag is that p is *not* now given by (13.2) if $\mathscr{E}$ is the average (continuum) internal field because, as already discussed, this does not take account of the atomic structure of the material. Since it is a macroscopic average over the whole of the interior of the medium $\mathscr{E}$ includes points between atomic sites as well as those at atomic sites. $\mathscr{E}$ also includes the field due to the dipole we are actually considering, whereas we need to know the *local* field *acting on* a dipole within the material, i.e. $\mathscr{E}$ *minus* the field due to the dipole itself.†

13.10. The internal field

The first treatment of this problem was due to Lorentz who suggested that the dielectric around a particular dipole should be divided into two parts: (1) a sphere, centred on the dipole, having a radius of many atomic spacings and (2) the region outside the sphere (Fig. 13.5(a)). It can be shown for (1) that with any cubic arrangement of atoms the field at the dipole due to the atoms in the sphere around it is zero. The region beyond the sphere (2) is sufficiently large and distant that it may be treated as a continuum with a

† The effect of the dipole *on* neighbouring atoms should also be eliminated. We shall not consider this further. It is mainly significant in dealing with orientational polarization in liquids (Onsager).

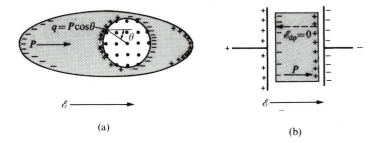

FIG. 13.5. (a) Calculation of the internal field on an atom. This is split up into the field due to the individual atoms inside an arbitrary sphere plus the field due to the rest of the dielectric outside the sphere (see text). (b) For the dielectric within a parallel-plate condenser there is no depolarizing field because the charges on the surface of the dielectric induce further opposite charges on the condenser plates.

polarization P. This will produce a charge density on the surface of the sphere similar to that discussed in section 13.5, which will have a value $P \cos \theta$, where θ is the angle between P and the normal to the spherical surface (Fig. 13.5(a)). By integrating over the surface the effective field at the centre of the sphere due to the surface charge can be shown to be $P/3\varepsilon_0$. This is called the Lorentz field. The local field $\mathscr{E}_{loc}$ acting on the dipole is therefore equal to the Lorentz field plus the external and depolarizing fields. For the simple geometry of a dielectric between the plates of a condenser the depolarizing field is zero (Fig. 13.5(b)). We then have

$$\mathscr{E}_{loc} = \mathscr{E}_{ext} + P/(3\varepsilon_0).$$

But from (13.2)

$$P = N\alpha\mathscr{E}_{loc}.$$

We can also define P in terms of the dielectric constant ε from eqns (13.7) and (13.8),

$$P = \varepsilon_0 \mathscr{E}_{ext}(\varepsilon - 1), \tag{13.11}$$

and by straightforward manipulation we can obtain the expression

$$\frac{N\alpha}{3\varepsilon_0} = \frac{\varepsilon - 1}{\varepsilon + 2}. \tag{13.12}$$

This is the Clausius–Mossotti formula. It shows the relationship between the atomic polarizability α and the macroscopic value of the dielectric constant ε. It has been adequately verified by taking measurements on gases at various pressures in order to change N.

Since, however, as we have already said, α cannot be calculated from theory, it is not possible to check the experimental values of α with any theoretical prediction.

When more than one polarization mechanism (section 13.3) is operating each will give a contribution to ε. If any of these cease to be operative it will be detected as a drop in ε. This is most marked in the case of polar liquids which have high values of ε due to orientational polarization. When they become solid ε drops dramatically because the rotational motion is frozen out.†

13.11. The frequency dependence of the dielectric constant

From the practical viewpoint the main interest in the theory of dielectrics is due to the fact that (1) the value of the dielectric constant varies with the frequency of the applied electric field and (2) that at certain frequencies energy can be dissipated within the dielectric. This is evident, even at fairly low frequencies, in 'lossy' condensers and at high frequencies by the dispersion of light and by the absorption of electromagnetic radiation within certain frequency ranges.

13.12. The relation between the refractive index and the dielectric constant

The development of Maxwell's equations for the electromagnetic field shows that the velocity c_m of electromagnetic radiation in a medium is equal to $(\mu\mu_0\varepsilon\varepsilon_0)^{-\frac{1}{2}}$. In free space μ and ε are unity, and so $c = (\mu_0\varepsilon_0)^{-\frac{1}{2}}$. If a medium is not magnetic so that $\mu = 1$ then the ratio of the velocities c/c_m is $\varepsilon^{\frac{1}{2}}$. But this ratio is the definition of the refractive index, n, and hence by experiment we should be able to confirm the relationship $n = \varepsilon^{\frac{1}{2}}$.

Measurements show that in general this expression does not seem to be valid. ε is found to be much greater than n^2. Thus NaCl has a low-frequency value of ε of 5·6, whereas n^2 is 2·25. Water with its polar molecules is even more extreme—ε is ~ 80 whereas n^2 is 1·75. Only in a material such as diamond in which the sole polarization mechanism is induced electron polarization is there very good agreement—ε is 5·68 and n^2 is 5·66. The reason for these discrepancies is quite straightforward. The refractive index is determined at optical frequencies ($\sim 10^{15}$ Hz), whereas the dielectric constant is usually measured at much lower frequencies. If ε is measured as a function of frequency it is usually found to decrease, not gradually, but by a series of fairly sharp drops at certain frequencies, until it approaches n^2 (Fig. 13.6). Associated with each of these drops there is a region of energy dissipation or dielectric loss.

---

† Some polar solids have a high value of ε which is only quenched at a temperature which is substantially *below* the freezing point. It has been suggested that in some cases this high value of ε is *not* due to rotational polarization in the solid but to a disordering process, whereby the dipoles undergo a transition from an antiferroelectric state at low temperatures to a para-electric state above the transition.

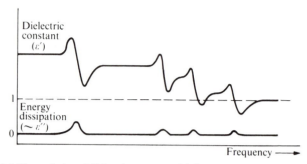

FIG. 13.6. The variation of dielectric constant with frequency. ε' drops in steps and at high frequencies it approaches the square of the refractive index. ε'' has a peak at each step.

These rapid changes in dielectric constant suggest that for each of the frequencies at which they occur, one of the polarization mechanisms 'switches off', because the polarization can no longer keep in step with the applied field. There are two different types of mechanism which can give rise to this kind of behaviour—resonance absorption and dipole relaxation.

13.13. Resonance absorption

Any type of induced dipole can be imagined to consist of a positive and a negative charge (the simplest case being the nucleus and the electron charge cloud around it) which are bound together by elastic restoring forces. This system will have a natural (angular) frequency of oscillation ω_0.

If the system is driven by an alternating electric field at a frequency ω, it will not be able to follow this field properly if $\omega > \omega_0$. The system is analogous to the mechanical problem of the forced vibration of a simple harmonic system and although we give the calculation in classical terms, quantum mechanics makes no essential difference to the result. The problem is very simple to express mathematically. If x is the relative separation of the charges on a dipole then we may write

$$m\frac{\mathrm{d}^2x}{\mathrm{d}t^2} + b\frac{\mathrm{d}x}{\mathrm{d}t} + \omega_0^2 x = -e\mathscr{E}\exp(-\mathrm{i}\omega t), \qquad (13.13)$$

where m is the mass, say, of the electron and $\mathscr{E}$ is the amplitude of the electric field. Note that in (13.13) we have included a damping term with a coefficient b which is proportional to the velocity of the particle, because there will always be some interaction mechanism for energy dissipation.

The steady state solution of (13.13) (neglecting transient effects) is standard:

$$x = \frac{-e\mathscr{E}}{m\{(\omega_0^2 - \omega^2) - \mathrm{i}b\omega\}}\exp(-\mathrm{i}\omega t) \equiv x_0\exp(-\mathrm{i}\omega t). \qquad (13.14)$$

Now x, the displacement of the charge, is proportional to the polarizability α, and hence α will also vary as (13.14). Since the amplitude of x is complex it is more convenient to separate it into real and imaginary parts as

$$x_0 = \frac{-e\mathscr{E}}{m}\left[\frac{(\omega_0^2-\omega^2)}{\{(\omega_0^2-\omega^2)^2+b^2\omega^2\}}+\frac{ib\omega}{\{(\omega_0^2-\omega^2)^2+b^2\omega^2\}}\right]. \quad (13.15)$$

The dielectric constant will have a similar form to (13.15) and it can be written as $\varepsilon = \varepsilon' + i\varepsilon''$, where ε' and ε'' are the real and the imaginary parts of ε respectively.

The two terms in the bracket of (13.15) are drawn in Fig. 13.7. We see that the real part of the function has a high value at low frequencies and it drops to a much lower value at the resonant frequency ω_0. For frequencies well away from ω_0, ε' is still changing; note that there is a minimum and a maximum on either side of ω_0. We therefore see that this type of resonance interaction could account for the frequency dependence shown in Fig. 13.6.

A change of ε with frequency implies that the refractive index is also frequency dependent and this type of dispersion is of course well known in all optical materials.

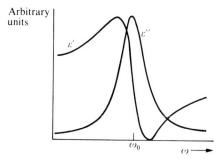

FIG. 13.7. Resonance absorption. The variation of ε' and ε'' as a function of frequency in the neighbourhood of an absorption line.

13.14. Energy dissipation

What is the physical significance of the imaginary term in (13.15)? We see that it is peaked at ω_0 and it has a line width which is proportional to the damping coefficient b.† Its importance, as we shall show, lies in the fact that it gives rise to energy dissipation.

The power W, dissipated in a system carrying a current density J in an electric field $\mathscr{E}$ is

$$W = J\mathscr{E}.$$

† This can be shown by straightforward algebraic manipulation.

In a dielectric the only current which can flow is the Maxwell displacement current dD/dt. Thus from (13.9) J will be

$$J = dD/dt = (d/dt)\{\varepsilon'\varepsilon_0\mathscr{E} \exp(-i\omega t) + i\varepsilon''\varepsilon_0\mathscr{E} \exp(-i\omega t)\}$$
$$= -i\omega\varepsilon'\varepsilon_0\mathscr{E} \exp(-i\omega t) + \omega\varepsilon''\varepsilon_0\mathscr{E} \exp(-i\omega t).$$

On taking the product of the real part of J with the real part of $\mathscr{E}$, and averaging over one cycle, the power dissipated may be shown to be

$$\tfrac{1}{2}\varepsilon_0\varepsilon''\omega\mathscr{E}^2. \tag{13.16}$$

Thus the dielectric loss is proportional to the imaginary part of the dielectric constant and it will occur at a frequency centred on ω_0.

We have discussed resonance absorption in terms of the vibrations of a very simple dipole system. If, however, we have an ionic crystal there will be resonances associated with the mutual displacement of the positive and negative charges and in more complicated organic materials, such as polymers, various parts of the molecules may have different resonant frequencies and so the frequency dependence of the dielectric constant of these materials can show a complicated structure. Ultimately at a sufficiently high frequency the only polarization mechanism which remains is the induced atomic polarization. This usually has a resonance at 10^{15}–10^{16} Hz so that for all materials the dielectric constant (and the refractive index) drops to a low value in the ultraviolet and there is a strong absorption—but the materials become transparent again as the X-ray region is approached.

13.15. Dielectric relaxation

The other effect which can give rise to a frequency dependent dielectric constant apparently arises only with polar molecules. If a permanent dipole is oriented in an electric field and it is then displaced, it will vibrate about the field direction and eventually by interaction with its surroundings it will *relax* back to its original position. The rate of energy transfer to the surroundings may be assumed to be of the form $\exp(-t/\tau)$ where τ is some characteristic time, the relaxation time, which is a measure of the strength of the interaction between the dipole and the system to which it is transferring energy—in our case it will be the phonon system. It should be noted that this relaxation time has no direct connection with the natural period of oscillation of the system discussed in the preceding section.

If the period of oscillation of an applied field is of the order or higher than τ, we shall again have the problem that the dipole will not be able to follow the field. The theory was first developed by Debye and it is beyond the scope of this book but as in the case of resonance absorption it results in a complex polarization and hence a complex dielectric constant. The Debye expressions are as follows:

$$\varepsilon' = \varepsilon_\infty + \frac{\varepsilon_s - \varepsilon_\infty}{1 + \omega^2 \tau^2}, \tag{13.17}$$

$$\varepsilon'' = \frac{(\varepsilon_s - \varepsilon_\infty)\omega\tau}{1 + \omega^2 \tau^2}, \tag{13.18}$$

where ε_s and ε_∞ are the low frequency (static) and very high frequency dielectric constants respectively. The curves are drawn in Fig. 13.8. The transition from high to low ε occurs at $\omega_0 = \tau^{-1}$ and it tends to be broader than that for resonance absorption. It should be noted that, unlike resonance absorption, ε' has no maximum and minimum on either side of the drop at ω_0. There is, however, a strong energy absorption term in ε'' and it is peaked at τ^{-1}.

Nearly all the data on dielectric relaxation have been obtained for liquids and gases. In solids, permanent dipoles seem to be too tightly bound to be able to rotate and it is not altogether certain whether any of the observed variations in the dielectric constant are a result of simple relaxation mechanisms. In any case it is difficult to distinguish between effects due to highly damped resonances and hindered rotation.

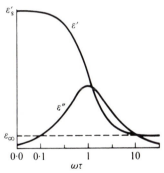

FIG. 13.8. Relaxation absorption. The variation of ε' and ε'' as a function of frequency $(\omega\tau)$ (eqns 13.17, 13.18).

13.16. Crystal lattices with a low symmetry

The simple crystal lattices which we have been happy to draw in this book have usually been adequate to explain the basic facts of solid state physics. They have all been pleasantly symmetrical and agreeable to view. But we now wish to describe briefly two sets of phenomena which are dependent on a crystal lattice having a low symmetry and hence a rather complicated structure. For this reason we have an even better excuse than usual not to go into details.

If a crystal has a structure in which the positions of the positive ions do not have a simple symmetric disposition to those of the negative ions the material will have a net electric dipole moment. This can be illustrated by drawing interpenetrating positive and negative cubic sub-lattices which are offset with respect to one another as is shown in Fig. 13.9. This type of situation can be achieved in two ways—by letting nature do the work so that the crystal spontaneously changes to a low symmetry structure. This is what happens in ferroelectrics. Or we can do the work by stressing the crystal, in which case we can produce piezoelectricity.

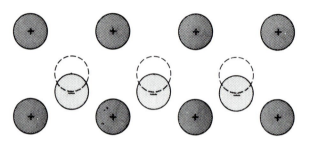

Fig. 13.9. An assymetric lattice can have a permanent dipole moment. This can be produced by an external force (piezoelectricity) or by internal atomic re-arrangement (ferroelectricity).

13.17. Ferroelectricity

Below a certain temperature it is found that some materials spontaneously acquire an electric dipole moment. By analogy with the magnetic case these materials are called ferroelectrics. Just as with ferromagnets these crystals exhibit a hysteresis curve of P versus $\mathscr{E}$ (Fig. 13.10) and this can be explained

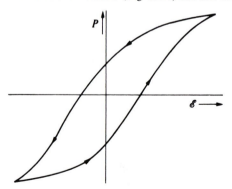

Fig. 13.10. A ferroelectric hysteresis curve.

by a domain hypothesis (sections 12.7, 12.10). These domains are quite easy to observe with polarized light in some materials (Fig. 13.11).

The transition to the ferroelectric state is a cooperative phenomenon which is accompanied by a specific heat anomaly or by a latent heat and it appears that at the transition temperature the crystal lattice spontaneously distorts to a more complicated structure which possesses a permanent electric dipole moment.

There are three main types of crystal structure which exhibit ferroelectricity: (1) the Rochelle salt structure, typified by Rochelle salt, $NaK(C_4H_4O_6) . 4H_2O$, (2) the perovskite group, consisting mainly of titanates and niobates, of which barium titanate, $BaTiO_3$, has been the most extensively studied, and (3) the dihydrogen phosphates and arsenates, e.g. KH_2PO_4 ('KDP').

The transition to an antiferroelectric state has also been detected in some materials. It consists of an ordered antiparallel array of dipoles and it can be detected by a sharp change in the dielectric constant at the transition temperature.

FIG. 13.11. Ferroelectric domains in gadolinium molybdate. Micrograph taken with crossed polarizers. (Magnification × 124.)

13.18. Piezoelectricity

If certain crystal structures are deformed by an external stress they can acquire a net dipole moment. This moment produces surface polarization charges which can be detected by measuring the potential developed across opposite faces of the crystal which acts as if it was a charged capacitor. For moderate stresses the polarization (and hence the potential) is proportional to the applied stress and so if the stress is reversed the potential changes sign.

These materials also exhibit the converse effect. If they are placed in an electric field they distort.†

It is fairly clear even from this very simple discussion that all materials which are ferroelectric will also exhibit piezoelectricity but the reverse is not true. A piezoelectric material must undergo a *spontaneous* distortion before it can become ferroelectric—and this only occurs in very special cases.

The piezoelectric effect is used in a wide variety of devices, e.g. gramophone pickups, microphones, and strain gauges. Barium titanate, which can be made as a ceramic and so can be formed into complicated shapes, is particularly useful. The converse effect is used in ultrasonic generators and a mechanically resonant piezoelectric crystal (always quartz) is at the heart of all fixed-frequency oscillators of high stability.

13.19. Dielectric breakdown

The very small currents that will usually be flowing within a dielectric can increase very rapidly if the applied field exceeds a certain critical value. This is called dielectric breakdown. The material conducts electricity and this limits the operating conditions for dielectrics which are used as insulators or in capacitors. The breakdown can be a reversible process so that when the field is reduced the dielectric properties are unimpaired but in other cases irreversible changes can occur and a conducting metallic or carbon track can be formed in the path of the breakdown current.

There are three mechanisms which can give rise to breakdown in solids:

(1) Electron impact ionization. In very high fields the few conduction electrons which will always be present are accelerated to such high energies that on collision with atoms they can dislodge an electron (this is another way of saying that they can excite an electron from the valence to the conduction band). This electron can also be accelerated to collide with another atom, and so an avalanche effect can build up which results in a very rapid increase in the current.

(2) If the dielectric losses are high (i.e. at frequencies in the neighbourhood of a relaxation or a resonance peak) the heat that is generated within the dielectric might exceed the rate at which it can leak away by thermal conduction. The material can then be raised to such a high temperature that it can conduct.

(3) Certain materials, e.g. mica and ceramics, can contain small inclusions of gas and in high fields arcing can occur through these regions.

† The converse piezoelectric effect should *not* be referred to as electrostriction. The term electrostriction is reserved for the phenomenon, which is shown by *all* materials of a dimensional change in an electric field $\mathscr{E}$. It is proportional to $\mathscr{E}^2$ and it is a very small effect.

13.20. Long-term effects in dielectrics

In conclusion we should note that most dielectrics contain a small amount of almost immobile extrinsic charge due to trapped ions, lattice vacancies, etc. These charges often give rise to slowly varying changes even in a static field. They are responsible, for example, for the fact that a charged condenser, after being short-circuited, often regains some charge on being left for several minutes. It is therefore usually impossible to obtain a reliable value for the d.c. dielectric constant by purely d.c. methods; instead a low frequency (~ 50 Hz) a.c. measurement should be made.

PROBLEMS

13.1. Assuming that an atom consists of a uniform sphere of negative charge with radius R surrounding a point positive charge, show that the polarizability is equal to $4\pi\varepsilon_0 R^3$. (The negative charge may be taken to remain uniform in an applied field.) The diameter of an argon atom is 0·3 nm. Estimate the refractive index of gaseous argon at S.T.P.

13.2. The dielectric constant of a solid is 5. It is placed between the plates of a condenser which are 1 mm apart and which is charged to 100 V. Calculate the local field acting on an atom in the dielectric.

13.3. A spherical cavity of radius R is cut in an infinite dielectric which has dielectric constant ε. A very small conducting sphere of radius r is placed at the centre of the cavity and a uniform electric field E_0 is set up in the main body of the dielectric. Calculate the induced moment of the sphere.

13.4. KH_2PO_4 is a colourless crystalline material which has a relative dielectric constant at low frequencies of 100. On the assumption that the low-frequency dielectric constant is due predominantly to the vibration of the H^+ ions (of which there are 2×10^{28} per cubic metre) about their equilibrium positions, estimate the frequency at which the peak of the dielectric absorption will occur.

13.5. Show that in a material which has a dielectric relaxation mechanism the dielectric power loss at very high frequencies is independent of the frequency.

13.6. If the dielectric relaxation time of a medium is 10^{-10} s estimate the width of the relaxation peak between points where the energy dissipation is reduced by one half.

13.7. The amplitude of a light wave which is travelling in a medium can be represented by $A \exp\{i(n\omega x/c - \omega t)\}$, where n is the refractive index and c is the velocity of light *in vacuo*. Show that if the light is attenuated as it passes through the medium then this can be represented by taking n as complex, i.e. in the form $n + ik$. Express ε' and ε'' in terms of n and k.

13.8. A medium which exhibits resonance absorption has a single absorption line centred on 600 nm (*in vacuo*). If the intensity of a beam of light of that wavelength falls to $\exp(-1)$ of its initial value in 0·025 m, find the maximum value of the imaginary part of the refractive index.

14. Superconductivity

14.1. Introduction

THE phenomenon of superconductivity—the disappearance of the d.c. electrical resistivity of certain metals at low temperatures—is surely one of the most unusual phenomena of solid-state physics. The vanishing of the resistance in itself is surprising enough, but since it is associated with other equally remarkable effects, superconductivity has been the source of a wealth of experimental and theoretical work, much of which can only be briefly touched upon in this chapter.

14.2. Zero resistivity

The most outstanding property of a superconductor is, of course, the complete disappearance of the electrical resistivity at some *critical temperature*, T_c, which is characteristic of the metal. So far about twenty-five elements and many thousands of alloys and compounds have been shown to be superconducting. The superconducting elements with their respective values of T_c are indicated in the periodic table shown in Fig. 14.1. A study of this table gives no strong clue which might help to explain the phenomenon, although metals which have a relatively simple quasi-free electron system, i.e. the alkalis and the noble metals, do not become superconducting. Elements representing most types of crystal structure can be superconductors. The highest value of T_c for an element is about 9 K, for niobium, followed closely by lead at 7·22 K. The lowest value of T_c yet reported is that for tungsten at 0·01 K. The highest value of T_c so far found is over 90 K, for some complicated ceramic type materials of the form Y–Ba–CuO.‡

The actual value of T_c depends very little on the particular sample of the metal which is measured,† although the width of the temperature range over which the resistance drops to zero is much narrower for pure, single-crystal specimens. Thus for a high purity, single crystal of tin, the transition region is 10^{-3} K wide, whereas it can be 10^{-1} K or even more for an impure or a strained sample.

Is the resistance in the superconducting state really zero or is it just very small? There have now been several investigations in which a current has been started in a closed superconducting ring by magnetic induction. It has been found that, providing the ring is kept below T_c, the *persistent current*, as it is called, does not diminish in value within the limits of measurement, even for a period of over a year. From these observations it appears that the upper limit

---

† Unless it contains magnetic impurities, when even small traces can depress T_c very considerably.

‡ This chapter is being revised just at the time when these new high T_c materials are being produced. The discovery of these materials is the most exciting development in solid state physics in recent years.

1	2	3	4	5	6	7	8	9	10	11	12	13	14	15	16	17	18
1H																	2He
3Li	4Be											5B	6C	7N	8O	9F	10Ne
11Na	12Mg											13Al 1·20	14Si	15P	16S	17Cl	18A
19K	20Ca	21Sc	22Ti 0·39	23V 5·3	24Cr	25Mn	26Fe	27Co	28Ni	29Cu	30Zn 0·88	31Ga 1·09	32Ge	33As	34Se	35Br	36Kr
37Rb	38Sr	39Y	40Zr 0·6	41Nb 9·3	42Mo 0·95	43Tc 8	44Ru 0·49	45Rh	46Pd	47Ag	48Cd 0·56	49In 3·4	50Sn 3·7	51Sb	52Te	53I	54Xe
55Cs	56Ba	57La α4·9 β6 58–71	72Hf 0·1	73Ta 4·5	74W 0·01	75Re 1·7	76Os 0·66	77Ir 0·14	78Pt	79Au	80Hg α4·15 β3·95	81Tl 2·4	82Pb 7·2	83Bi	84Po	85At	86Rn
87Fr	88Ra	89Ac	90Th 1·37	91Pa 1·4	92U α0·68 β1·80												

FIG. 14.1. The periodic table showing the superconducting elements. The transition temperature for each superconductor is shown in the lower part of each box. The rare earth elements (58–71) and the transuranics (> 92) are not superconducting and are omitted. α, β, refer to different crystalline phases of the same element.

to the possible value of the resistance is $<10^{-25}$ ohm m, although of course to prove that it was actually zero would require an experiment extending to infinite time!

It should, however, be noted that the zero resistance only applies to the d.c. behaviour of a superconductor. For a.c., as we shall see, a superconductor does exhibit a resistance, although at low frequencies it is very small.

14.3. The critical field

The second fundamental property of a superconductor, besides its zero resistance, is that its normal resistance may be restored if a magnetic field greater than a critical value, B_c, is applied to the specimen.[†] B_c depends both on the material and on the temperature. It is zero at T_c and as the temperature is reduced it increases, following, approximately, a parabolic law of the form

$$B_c = B_0\{1-(T/T_c)^2\}, \qquad (14.1)$$

and hence it tends to a constant value B_0 as T approaches 0 K (Fig. 14.2). In general the higher the value of T_c, the higher will be B_0. For most pure metals

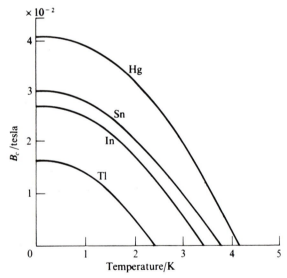

FIG. 14.2. Curves showing the dependence of the critical field, B_c, on the temperature for Tl, In, Sn, and Hg. (After Shoenberg, 1952.)

† B or H? Whilst many texts work in terms of the external magnetic intensity H, it is more helpful to develop the theory in terms of $B(=\mu_0 H)$, since it is the *field*, B, applied to a superconductor, which tends to be the most interesting and useful quantity.

B_0 is usually quite low. If T_c is around 1 K, B_0 is about 10^{-2} T (100 gauss), whereas for higher values of T_c it might range up to $\sim 10^{-1}$ T.

14.4. The critical current density

Associated with an electric current is a magnetic field and hence if a superconductor carries a current such that the field which it produces (plus any other external field) is equal to B_c, then the resistance of the sample will be restored. The current density at which this occurs is called the *critical current density*.

14.5. $B = 0$ and the Meissner effect

If a field B_{ext}, which is less than B_c, is applied to a superconductor a persistent current will be induced in a direction which will oppose the effect of the applied field (Lenz's law). The field generated by this current will exactly cancel B_{ext} and, since it does not die away, the effective magnetic field within the superconductor, B_{int}, remains zero. The induced persistent current is restricted to a surface layer of finite depth in which the external field is not exactly cancelled but inside that layer $B_{int} = 0$ (Fig. 14.3(b)). If the external field is increased (path AB_cC, Fig. 14.3(a)) then when it reaches B_c the persistent currents will die out and B_{int} will be equal to B_{ext} (Fig. 14.3(c)). If B_{ext} is now reduced (path CB_cA) then as it passes B_c again, persistent currents are once more set up so that $B_{int} = 0$ as in the initial state, and the situation will return to that in Fig. 14.3(b). Whilst this behaviour is not, perhaps, surprising, the expulsion of the magnetic field when B_{ext} is reduced below B_c is not a direct consequence of the zero resistivity of a superconductor. It is a fundamental additional property and it is called the *Meissner effect*. It may also be demonstrated by cooling a superconductor which is initially in a field B_{ext} above T_c until it is at a temperature T for which B_{ext} is equal to B_c (path DB_cE). At that point the field will be expelled so that $B_{int} = 0$ as in Fig. 14.3(b).

A further interesting effect occurs if the sample is a hollow cylinder or a torus. On either path, CB_cA or DB_cE, the field is expelled from the material at B_c but it is *trapped* in the central hole (Fig. 14.3(d)).

For the simple case of a bulk superconductor, when $B_{int} = 0$ we may use the following simple analysis. From standard electromagnetic theory

$$B_{int} = \mu_0 H + \mu_0 m_v$$

$$= B_{ext} + \mu_0 m_v,$$

where m_v is the magnetic moment per unit volume. Since $B_{int} = 0$ we then have

$$m_v = -B_{ext}/\mu_0. \tag{14.2}$$

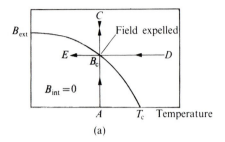

(a)

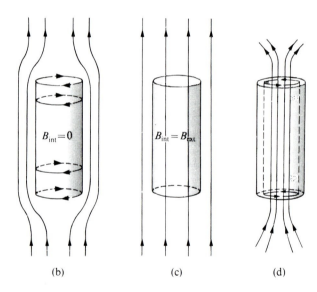

(b)   (c)   (d)

FIG. 14.3. $B_{int} = 0$ and the Meissner effect. (a) Phase diagram showing the region (shaded) where $B_{int} = 0$. The field is expelled at B_c on the paths CB_cA and DB_cE. (b) $B_{int} = 0$ due to the cancellation of B_{ext} by the field generated by the circulating currents on the surface of the specimen. (c) $B_{int} = B_{ext}$ when $B_{ext} > B_c$. (d) The magnetic field within a hollow cylinder is trapped when it becomes superconducting. It is produced by the circulating currents as indicated.

The superconductor therefore acts as if it possesses a magnetic moment (Fig. 14.4(a)) which opposes the external field and it is sometimes said to be a

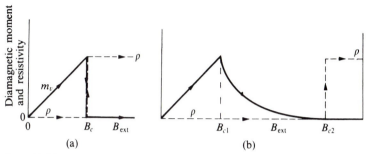

FIG. 14.4. The destruction of superconductivity by a magnetic field, B_{ext}. (a) In type-I materials there is a sharp transition at $B_{ext} = B_c$ and there the diamagnetic moment, m_v, drops to zero and the resistivity, ρ, rises to its normal value. (b) In type-II materials the field begins to penetrate at $B_{ext} = B_{c1}$ and m_v starts to decrease, but the resistance does not return. The transition is not complete until $B_{ext} = B_{c2}$.

perfect diamagnet. It should be noted, however, that this diamagnetic effect is due to macroscopic currents circulating on the surface of the material; it is not caused by any microscopic or atomic mechanism.

14.6. Type-I and type-II superconductors

Superconductors may be divided into two classes which depend on the way in which the transition from the superconducting to the normal state proceeds when the applied field exceeds B_c. Let us here consider the simple geometry when B_{ext} is aligned parallel to the axis of a long cylindrical specimen. In type I materials, as B_c is reached, the entire specimen enters the normal state practically simultaneously, the resistance returns, the diamagnetic moment becomes zero and $B_{int} = B_{ext}$ (Fig. 14.4(a)). Type-I behaviour is typical of almost all the elements.[†]

In type-II superconductors the transition to a completely normal specimen is much more gradual. At the critical field, which is now called B_{c1}, very fine filaments of the material become normal and a cylinder of magnetic flux—a flux quantum (section 14.23)—is centred on each filament. This is called the *mixed state*. As the external field is further increased more normal filaments are produced, each supporting a flux line, and the diamagnetic moment decreases. Only when a field B_{c2} is attained does the entire specimen become completely composed of normal filaments and it is only then that the electrical

---

[†] With a different field geometry type-I materials exhibit the *intermediate state*. This is described in section 14.12.

resistance returns (Fig. 14.4(b)). B_{c1} and B_{c2} are called the lower and upper critical fields, respectively.[†] The filamentary structure is shown in Fig. 14.5.

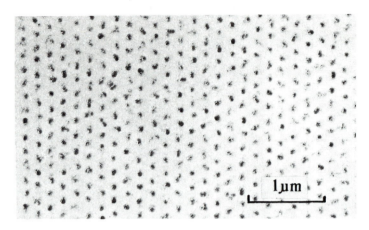

FIG. 14.5. The filamentary structure of the mixed state of a type-II superconductor (niobium). The dark spots are normal regions on which ferromagnetic powder has been deposited. (Photo by courtesy of Dr. U. Essmann).

Type-II behaviour is typical of most alloys and also of the elements Nb, Tc, and V. The superconducting materials used in high-field magnets are all type II, but, as we shall see, other properties, apart from being type II, are necessary in order that superconductors may be useful in high-current and high-field technology. Most type-II materials are *not* useful for these applications.

The division of superconductors into the two types is not just an empirical one. There are sound theoretical reasons for this distinction and these will be discussed in section 14.19.

14.7. What causes superconductivity?

As we saw in Chapter 8, the resistance of metals at low temperatures is caused by the scattering of the conduction electrons by impurity atoms. In no way can the quasi-free electron model be modified so that the interaction with impurities can be reduced to zero. These impurities are always present and they will always scatter the electrons. A completely fresh approach to the problem is necessary in order to explain the disappearance of the resistivity.

The fundamental idea underlying the modern theory of superconductivity is that the electrons *pair up* with one another due to a special type of attractive

† Beyond B_{c2} there is another critical field B_{c3}, which is that needed to destroy the superconductivity associated with the surface states of the material, but we shall not discuss this phenomenon here.

interaction. The zero resistivity is then accounted for by assuming that a pair can only be scattered if the energy involved is sufficient to break it up into two single electrons. In general this energy will not be available and so the electron pair passes on, undeviated by impurities.

Of course the ordinary Coulomb interaction between electrons produces a repulsion, and the attractive pair interaction is much more subtle. It is an indirect interaction and it is caused by the way a positive ion in the crystal responds to the passage of electrons in its vicinity. Let us consider an electron passing close to an ion. There will be a momentary attraction between them which might slightly modify the vibration of the ion. This in turn could interact with a second electron nearby which will also be attracted to the ion (Fig. 14.6); but the net effect of these two interactions is that there is an apparent

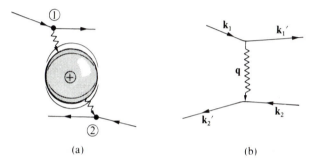

(a)   (b)

FIG. 14.6. (a) The pairing interaction between electrons occurs because the motion of electron 1 modifies the vibration of the ion and this in turn interacts with electron 2. The over-all effect can be a net attractive force between the electrons. It should be noted that, although the two electrons are here shown fairly close together, in reality they can be a coherence length apart (section 14.17)—this may be many *thousand* atomic spacings! (b) The pairing can be considered to be due to the emission of a phonon of wave vector q by electron 1, followed by its absorption by electron 2 (eqn (14.3)). k, k' are the electron wave vectors before and after the interaction.

attractive force between the two electrons and this would not have arisen if the ion had not been present. In the language of field theory the interaction is said to be due to the exchange of a *virtual phonon, q,* between the two electrons. In terms of the wave vectors, k, of the electrons 1 and 2, the process can be formally written as

$$k_1 - q = k'_1 \quad \text{and} \quad k_2 + q = k'_2; \tag{14.3}$$

hence $k_1 + k_2 = k'_1 + k'_2$, and so the net wave vector of the pair is conserved (Fig. 14.6(b)).

Now whilst this pairing interaction might in principle exist, it could be extremely weak. The important step was made by Cooper in 1956, who showed that if there was an attractive interaction then, however weak it was, the lowest energy state (the ground state) of the system at $T = 0$ K would be one in which the electrons were paired. The pairs are therefore sometimes called *Cooper pairs*.

The development of this pairing hypothesis culminated in the theory proposed by Bardeen, Cooper, and Schrieffer in 1957 (always referred to as the B.C.S. theory) and this is the basis of our current understanding of superconductivity.

The B.C.S. theory relies on advanced quantum mechanical techniques which treat the entire electron assembly as one coupled system. It would be quite out of place in this book to go into any detail. It should either be done properly or not at all, for no amount of handwaving can make up for omitting the quantum field theory. We shall therefore discuss the main results and achievements of the theory without giving any justification for them.

14.8. Further consequences of electron pairing

The pairing energy depends on the strength of the interaction between the electrons and the ions and since the energy involved is quite small, the pairs can be broken by thermal activation. The pairs will begin to form at the transition temperature, T_c. As the temperature is further reduced more pairs will be able to remain stable, until at 0 K all possible electron pairs will be formed. Thus even when the material is superconducting, there will always be some unpaired electrons (called *quasi-particles* or *normal* electrons) present. It is interesting to note that the idea of two types of electronic state was the basis of the *two-fluid* model of superconductivity (1934). This was a purely *ad hoc* hypothesis to account for the then known properties of superconductors. In that model the electron assembly was considered to be composed of two interpenetrating *fluids*, of superconducting and normal electrons† whose relative proportions varied with the temperature following the relation $\{1 - (T/T_c)^4\}$. The obvious analogy between the paired and unpaired electrons of the B.C.S. theory and the superconducting and normal fluids of the two-fluid model has meant that some of the older terminology and ideas have been retained within the framework of the B.C.S. theory.

Even before the B.C.S. theory certain experiments suggested that superconductivity was connected with some type of lattice interaction. It was known, for example that ordinary (white) tin is a superconductor whereas the other allotropic form, grey tin, is not; bismuth, which is not normally a superconductor, becomes superconducting at very high pressures (~ 20 kBar). Thus superconductivity is not a property of the *atom* itself but it depends on the crystal lattice arrangement.

---

† The two-fluid model gave no fundamental explanation to account for the properties of the superconducting fluid.

The most outstanding early experiments, however, were those which indicated a direct connexion with the lattice *vibrations*. These were observations of the *isotope effect*. They showed that values of T_c for samples of different isotopes of the same element are roughly proportional to $M^{-\frac{1}{2}}$, where M is the atomic mass† (Fig. 14.7). This may be explained in terms of the frequency of

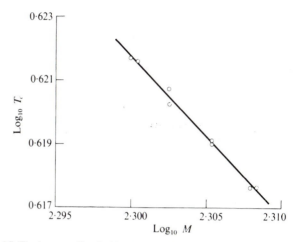

FIG. 14.7. The isotope effect in Hg. A logarithmic plot of the transition temperature against the atomic mass for several isotopically enriched samples. The line has a slope of -0.504, showing good agreement with the $M^{-\frac{1}{2}}$ prediction (Reynolds, Serin, and Nesbitt, 1951).

the ionic vibration. If the elastic constant is unchanged the ionic frequency for any simple oscillator will be proportional to $M^{-\frac{1}{2}}$. The isotope effect thus demonstrates very effectively that superconductivity is intimately connected with *phonon* interactions.

Since an electron pair has a lower energy than two normal electrons there is an *energy gap* between the paired and the two single-electron states. This energy is often denoted by 2Δ. Thus the net energy to excite *each* electron is Δ, although of course both must be excited at the same time. In principle any two electrons can pair, providing that their net wave vector is conserved before and after the exchange of the virtual phonon as in eqn (14.3). It can be shown, however, that when zero current is flowing, the most probable distribution and the one that corresponds to the superconducting ground state is the one in

† The power of M is not exactly $-\frac{1}{2}$. The B.C.S. theory shows that higher order terms can account for these deviations.

which the pairs consist of electrons of equal and opposite wave vectors, k, and opposite spins, i.e. $(+k\uparrow, -k\downarrow)$. All pairs have the same energy and the state can be considered as one of zero entropy. Since the net k of a pair is zero the pairs have no momentum and so this is not a current-carrying state. If a current is flowing the pairs are formed so that the net wave vector for all of them is the same, i.e. $\{(k+\delta)\uparrow, (-k+\delta)\downarrow\}$.

The energy gap, 2Δ, is *not* equal to the energy associated with a single virtual phonon interaction although it is intimately connected with it. The single interaction occurs at an instant and is then finished. For the attraction between a pair to be maintained there must be a continual exchange of virtual phonons and hence a *continuing* change of electron pair states. This demands a supply of *vacant* states such as k'_1 and k'_2 in eqn (14.3). If there are a large number of suitable states available the probability of virtual phonon interactions will be high and the pair binding energy (2Δ) will be strong. At temperatures close to T_c, however, there will be a large number of single electrons excited and the number of states available for pair interactions will be reduced. For since in the ground state $k'_1 + k'_2 = 0$, if a *single* electron is in state $+k'$ then it blocks a pair interaction into states $(+k', -k')$. Thus close to T_c 2Δ decreases rapidly and at T_c it becomes zero (Fig. 14.8). At 0 K the B.C.S.

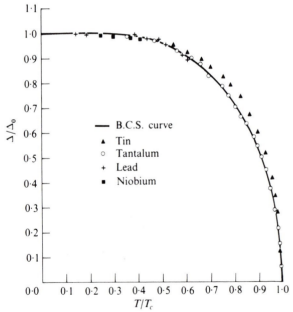

FIG. 14.8. The temperature variation of the energy gap compared with its value at 0 K for Sn, Ta, Pb, and Nb. The full line is calculated from the B.C.S. theory. These are results from electron tunnelling experiments (section 14.21) (Townsend and Sutton, 1962).

theory predicts that 2Δ should be about $3.5kT_c$ and, as we shall see, this is borne out quite well by experiment.

14.9. The thermodynamics of a superconductor

It is very straightforward to derive expressions which relate the entropy and the specific heat differences between the normal and the superconducting states with the temperature variation of B_c. In a reversible change† the transition from one phase to another occurs when the Gibbs' free energy, G, is the same in each phase. If the substance has a magnetic moment, m_v, per unit volume, we may write the free energy of the superconducting phase in a field $B\,(B < B_c)$ as

$$G_s = U - TS - V\int_0^B m_v\, dB,\tag{14.4}$$

where U is the internal energy and the other symbols have their usual meanings.‡

Consider the case of a long thin type-I specimen with its axis parallel to B and for which, as we have already described, the transition to the normal state occurs practically discontinuously at $B = B_c$. For $B < B_c, m_v$ is given by (14.2) and so, on integrating, eqn (14.4) becomes

$$G_s(B_c) = G_s(0) + \tfrac{1}{2}VB_c^2/\mu_0.\tag{14.5}$$

$G_s(0)$ is the free energy of the superconducting phase in zero field. $G_s(B_c)$ is the free energy at the transition and so it is the same for both the normal and the superconducting states at B_c. Hence $G_s(B_c) = G_n(B_c)$, the free energy of the normal phase. Thus the free energy difference between the two states is

$$\tfrac{1}{2}VB_c^2/\mu_0\tag{14.6}$$

Since the entropy, S, may be obtained from the relation

$$S = -\partial G/\partial T,$$

we obtain for the difference in the entropies between the two states

$$S_n - S_s = -(VB_c/\mu_0)(dB_c/dT).\tag{14.7}$$

---

† Many writers suggest that the justification for being able to use thermodynamics is that the superconducting transition is reversible. This is misleading. Thermodynamics can certainly be used for irreversible processes (e.g. the Joule–Kelvin effect). The criterion for the use of thermodynamics is that the system should have a *unique* equation of state.

‡ A justification of eqn (14.4) is given by A. B. Pippard (1957), *Elements of Classical Thermodynamics*, pp. 132–3, Cambridge University Press. Further discussion is given by F. N. H. Robinson (1973), *Macroscopic Electromagnetism*, pp. 174–6, Pergamon Press, Oxford.

From the curves of Fig. 14.2 we see that dB_c/dT is always negative and hence the entropy of the superconducting state is always lower than that of the normal state; i.e. the superconducting state is one of greater order. We also note that since, according to the third law of thermodynamics, $S_n - S_s$ must vanish at 0 K, so dB_c/dT must tend to zero at very low temperatures, and this trend is shown in Fig. 14.2.

From eqn (14.7) we see that, when superconductivity is destroyed by the application of a field greater than B_c, a latent heat, Q, equal to $T(S_n - S_s)$ will be absorbed, where

$$Q = -(VB_cT/\mu_0)(dB_c/dT). \tag{14.8}$$

There is, however, no latent heat associated with the onset of superconductivity at T_c (where $B_c = 0$).

14.10. The specific heat

The specific heat is $T\, \partial S/\partial T$, and hence the difference in the specific heats of the two phases, $c_s - c_n$, may be derived† from eqn (14.7) as

$$c_s - c_n = (T/\mu_0)\left\{B_c\frac{d^2B_c}{dT^2} + \left(\frac{dB_c}{dT}\right)^2\right\} \text{ per unit volume.}$$

Thus in zero field at T_c there is a discontinuity in the specific heat which is given by

$$c_s - c_n = (T_c/\mu_0)(dB_c/dT)^2_{T_c},$$

an essentially positive quantity. At lower temperatures $c_s - c_n$ becomes negative because (d^2B_c/dT^2) is always negative and (dB_c/dT) tends to zero as T approaches 0 K (Fig. 14.9(a)).

The specific heat in the superconducting state may be considered to consist of two components. When the temperature is raised from T to $T + dT$ thermal energy is needed, (1) to bring the system of unpaired (normal) electrons into thermal equilibrium at the new temperature $T + dT$, and (2) to excite more unpaired electrons from the paired states so that the ratio of paired to unpaired electrons is that which is appropriate to $T + dT$. In general, (2) is the more important term. We have already described that there is an energy gap, 2Δ, between the paired and the unpaired states, and so at very low T where 2Δ is constant we should expect that the number of unpaired electrons would be proportional to $\exp -(\Delta/kT)$.‡ Thus the number excited from the pair

---

† Both c_s and c_n will contain a lattice contribution to the specific heat, but this is assumed to be the same in both states. Hence $c_s - c_n$ should only be dependent on the difference in the electronic configurations of the two states.

‡ No, it is *not* proportional to $\exp -(2\Delta/kT)$! The Fermi energy lies in the middle of the gap and so the excited states lie Δ above E_F. The probability of occupation is therefore $1/\{\exp(\Delta/kT) + 1\} \approx \exp -(\Delta/kT)$ if $\Delta \gg kT$. The situation is similar to that in a semiconductor, see Fig. 9.7.

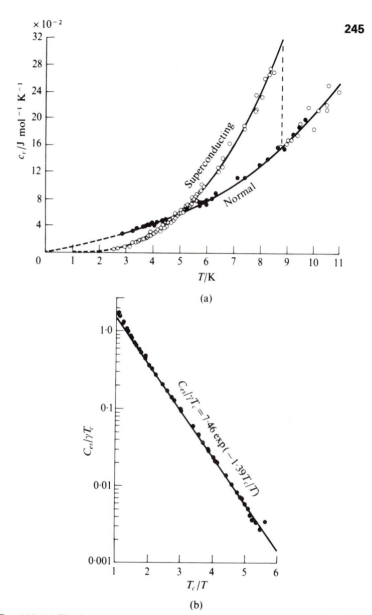

Fig. 14.9. (a) The heat capacity of Nb in the normal and superconducting states showing the sharp discontinuity at T_c (Brown, Zemansky, and Boorse, 1953). (b) The superconducting state electronic heat capacity of Ga as a function of temperature. This illustrates the exponential dependence of the heat capacity at low temperatures which is due to the excitation of paired electrons to single electron states across an energy gap (N. E. Phillips, 1964).

states will have the same proportionality and hence c_s should have an exponential temperature dependence. This is indeed observed, with a value of 2Δ of about $3 \cdot 5 \, kT_c$ and it is one of the important verifications of the B.C.S. theory (Fig. 14.9(b)).

14.11. The thermal conductivity

As already discussed in Chapter 8, the transport of heat in metals is due almost entirely to electron conduction—the lattice contribution is so small that it can nearly always be neglected. In a superconductor that is not always the case. Since the paired electrons cannot interact with the thermal phonons thermal energy cannot be transferred to the pairs and so they cannot transport heat. Only the unpaired electrons can contribute and since these decrease in number as the temperature is reduced, we would expect the thermal conductivity in the superconducting state to fall below that in the normal state and that this behaviour would be accentuated at lower temperatures. This is indeed the case (Fig. 14.10), but at a sufficiently low T another interesting effect is observed which is due to the phonon heat conduction. Lattice conduction in a metal is very small because the phonons are so strongly scattered by the electrons that the phonon mean free path is very short. In a superconductor, however, as we have just discussed, the phonons can only interact with the unpaired electrons and at very low temperatures there are so few of these that the phonon mean free path increases to such an extent that phonon conduction becomes the dominant mechanism for heat transport. The mean free path increases until it is limited by the specimen or the crystallite dimensions just as in a dielectric crystal (section 6.10) and the conductivity follows a T^3 dependence.

In the normal state the conductivity will be proportional to T, as in any metal (section 8.7) and so, since, as just discussed, the thermal conductivity in the superconducting state varies as T^3, it will be much less than in the normal state.

This behaviour has been put to good use. At very low temperatures a thin strip of superconductor (very often lead) is used as a *thermal switch* between the specimen and the heat sink. Thermal contact is achieved by applying a field greater than B_c, so that the link is in the normal state. The specimen may then be thermally isolated by removing the field. This technique avoids the vibration and heating which can be introduced by a mechanical device (see problem 14.4).

14.12. The destruction of superconductivity in type-I materials—the intermediate state

In the preceding sections we have only discussed the case in which the external magnetic field, B_{ext}, is applied in a direction which is parallel to the

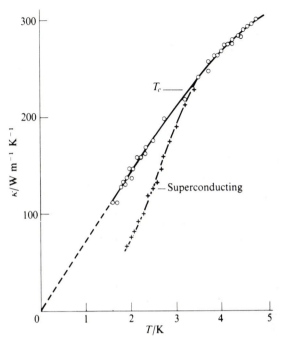

FIG. 14.10. The temperature dependence of the thermal conductivity of In, which illustrates the progressive reduction in the conductivity with decreasing T in the superconducting state (Hulm, 1950).

axis of a long thin type-1 specimen. The transition to the normal state is then at $B_{ext} = B_c$ and it is very sharp. This, however, is not so for other geometries. Let us now consider another simple example—that when B_{ext} is *perpendicular* to the axis of a long cylindrical specimen (Fig. 14.11(a)). Since B is zero within the specimen, the lines of force will be distorted so that they are concentrated more strongly at that part of the surface labelled AA' in the figure; i.e. the field at AA' will be greater than B_{ext}. The field is no longer uniform and it can be shown that at AA' it is actually twice B_{ext}. Thus when B_{ext} reaches $\frac{1}{2}B_c$ the field at AA' attains the critical value and parts of the specimen will go normal.†

---

† This is only for a cylindrical specimen. For other shapes the transition to the intermediate state occurs at different values of B_{ext}.

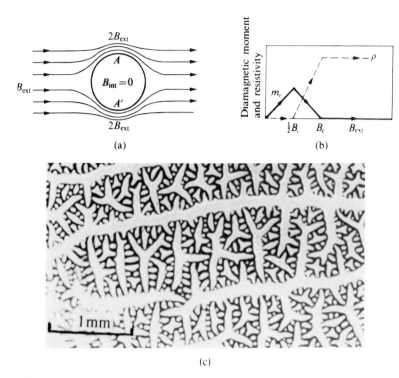

(a)   (b)

(c)

Fig. 14.11. The intermediate state. (a) When B_{ext} is perpendicular to the axis of a cylindrical specimen the lines of force are distorted so that the field at A and A' is $2B_{ext}$. Field penetration therefore starts at $B_{ext} = \frac{1}{2}B_c$. This is the onset of the intermediate state. (b) At $B_{ext} = \frac{1}{2}B_c$ the diamagnetic moment, m_r, starts to decrease and the resistivity, ρ, starts to return, but the transition is not complete until $B_{ext} = B_c$. Compare this behaviour with Fig. 14.4(a). (c) The structure of the intermediate state in lead. The normal regions are dark and are shown up by the deposition of Fe. Note that the scale is much coarser than that of Fig. 14.5 (Essmann and Träuble).

The specimen is then said to be in the *intermediate state*. As the field is increased further the normal regions grow until at $B_{ext} = B_c$ the entire specimen has become normal.

At the beginning of the intermediate state the electrical resistance along the specimen starts to return, thus showing that normal regions must extend right across it (Fig. 14.11(b)). When $B_{ext} = B_c$ the normal resistivity is completely restored.

The structure of the intermediate state has been investigated using several techniques which all rely on the fact that the field will be non-zero in the normal regions whilst remaining zero in the superconducting ones. Field-sensitive probes, magneto-optic effects and magnetic powder patterns all show an extremely complicated structure (Fig. 14.11(c)). The important observation, however, is that the various regions are of macroscopic dimensions—typically, say, 0·1 mm. As the field is increased the normal regions expand at the expense of the superconducting ones.

This behaviour of type-I superconductors in the intermediate state is dependent on the shape of the specimen and the direction of B_{ext}. It should be contrasted with the *mixed state* (section 14.6) which is an inherent feature of type-II materials and which occurs even for long specimens in longitudinal magnetic fields.

14.13. Field penetration; the London equations

We have already described that the zero value of B_{int} is a result of induced surface currents which themselves produce a field to oppose B_{ext} (section 14.5). These currents cannot flow in an infinitely thin surface layer because otherwise the critical current density (section 14.4) would be exceeded, and so they must extend into the material slightly. Only completely under this layer will the full influence of the surface currents be felt so that $B_{int} = 0$. In the layer itself the field will not be zero and hence an external field is not completely cancelled.

The Meissner effect and the field penetration cannot be deduced from Maxwell's electrodynamic equations alone. The brothers F. and H. London (1935) showed that it was necessary to introduce two additional equations in order to account for the magnetic properties. These equations, which cannot be rigorously justified, are

$$\text{curl } j_s = -B/\Lambda \tag{14.9}$$

and

$$\partial j_s/\partial t = \mathscr{E}/\Lambda, \tag{14.10}$$

where $\Lambda = m/(n_s e^2)$, m is the electron mass, n_s the number density of superconducting (i.e. paired) electrons, $\mathscr{E}$ is the electric field, and j_s is the current density of the superconducting (i.e. non-resistive) current.

The penetration of B can now be calculated as follows. We assume that we have quasi-static conditions so that $\partial j_s/\partial t = 0$ and therefore from (14.10) the electric field is zero. If this is so, there cannot be any flow of normal (unpaired) electrons and therefore the only current in the material must be superconducting current. We may therefore write the Maxwell equation in terms of j_s alone as

$$\text{curl } B = \mu_0 j_s.$$

Now take the curl of this expression

$$\text{curl curl } B = \mu_0 \text{ curl } j_s.$$

We substitute for curl j_s from the London equation (14.9)

$$\text{curl curl } B = -(\mu_0/\Lambda)B.$$

Recalling the vector relation curl curl = grad div $- \nabla^2$, and since div $B = 0$, we obtain

$$\nabla^2 B = (\mu_0/\Lambda)B. \tag{14.11}$$

In a very similar fashion we could start by taking the curl of (14.9) and obtain

$$\nabla^2 j_s = (\mu_0/\Lambda)j_s. \tag{14.12}$$

The solutions of (14.11) and (14.12) depend on the boundary conditions. The simplest case is that of a semi-infinite specimen which extends everywhere beyond $x = 0$. If B_{ext} is parallel to the surface in the z-direction then the solution to (14.11) for $x > 0$ is†

$$B_z = B_{ext} \exp -x(\mu_0/\Lambda)^{\frac{1}{2}}. \tag{14.13}$$

Thus the field falls off exponentially within the superconductor and this decrease is characterized by a parameter, the *penetration depth*, which is usually denoted by λ, where

$$\lambda = (\Lambda/\mu_0)^{\frac{1}{2}} \tag{14.14}$$

or, on substituting for Λ,

$$\lambda = (m/\mu_0 n_s e^2)^{\frac{1}{2}}. \tag{14.15}$$

If we assume that there is about one superconducting electron per atom so that n_s is about 10^{28} m^{-3}, then λ is of the order of 10^{-8} m. As the transition temperature is approached the number of paired electrons decreases and so λ will increase.

The solution of (14.12) for j_s will, of course, be similar to (14.13). j_s will be in the y-direction with the same exponential dependence on x.

14.14. Measurements of the penetration depth

Since the perfect diamagnetism of a superconductor (section 14.5) is dependent on B being zero throughout the material, the surface penetration of the field will reduce the diamagnetic effect. Thus measurements of the susceptibility can be used to investigate the behaviour of λ. A straightforward technique is to measure the value of a mutual inductance in which the primary and secondary windings are wound around the specimen. A typical set of results

---

† We have chosen the negative sign in the exponent of (14.13) since as x goes to infinity, B must tend to zero.

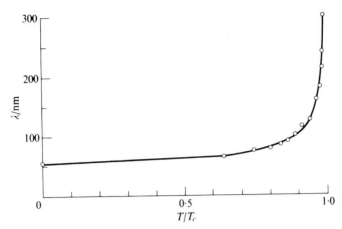

FIG. 14.12. The temperature variation of the penetration depth, λ, of a Sn single crystal which illustrates the rapid increase in λ as T_c is approached. (From the data of Schawlow and Devlin, 1959).

is shown in Fig. 14.12. The rapid rise of λ close to T_c, discussed in the previous section should be noted. The results indicate that λ has a form very close to

$$\lambda = \lambda_0 \{1 - (T/T_c)^4\}^{-\frac{1}{2}}$$

and by comparison with (14.15) this would suggest that the number of paired electrons varies as

$$\{1 - (T/T_c)^4\}. \tag{14.16}$$

The effect of field penetration will be enhanced if the specimen is prepared in the form of very small particles or fine wires or films, in which one or more dimensions are of the same order or smaller than λ. The diamagnetic effect will then be very much reduced. Many such experiments have been made[†] and they confirm the basic predictions of the London equations.

14.15. High-frequency effects

If an alternating potential is applied to a superconductor (14.10) shows that j_s will be $\pi/2$ out of phase with the electric field; i.e. the superconductor will have an *inductive reactance*. This is due to the inertia of the electrons. Although this reactance will not itself dissipate any energy, the field will also drive a

[†] See, for example, Shoenberg (1952), *Superconductivity* (2nd edn.), Chapter 5, Cambridge University Press.

current of *normal* electrons and this will be resistive.† This resistance will give rise to energy dissipation (see problem 14.11).

In order to discuss this resistive effect in more detail we should recall that at high frequencies the current in a normal metal flows only within a thin layer at the surface, the *skin depth, d*. This may be shown to be

$$d = (\tfrac{1}{2}\sigma\omega\mu\mu_0)^{-\frac{1}{2}}, \quad (14.17)$$

where ω is the angular frequency and σ is the electrical conductivity.‡ In the superconducting state the effective conductivity will be σ'. It will be proportional to the density of unpaired electrons and this, by (14.16) increases with T. Thus d will increase as T is reduced.

If we apply an electric field of the form $\mathscr{E} = \mathscr{E}_0 \exp(-i\omega t)$, then

$$j_n = \sigma'\mathscr{E}_0 \exp(-i\omega t), \quad (14.18)$$

and from (14.10)

$$j_s = -(i\omega\Lambda)^{-1}\mathscr{E}_0 \exp(-i\omega t), \quad (14.19)$$

and so the ratio of the moduli of the current densities will be

$$|j_n|/|j_s| = \sigma'\omega\Lambda.$$

After substituting for σ' from (14.18), for Λ from (14.14) and putting $\mu = 1$, we obtain

$$|j_n|/|j_s| = 2(\lambda/d)^2. \quad (14.20)$$

Thus the normal resistive current becomes comparable to or greater than the superconducting current when the skin depth becomes equal to or less than the penetration depth. The effect tends to become important at microwave frequencies and it increases as T_c is approached, due both to the increase in λ and the decrease in d. A typical set of results is shown in Fig. 14.13.

At high frequencies (usually in the far infra-red) superconductors retain their resistivity at *all* temperatures below T_c. This is also illustrated in the figure and it will be discussed in section 14.21.

14.16. Penetration depth measurements at high frequencies

Very accurate measurements of λ can be achieved by making the superconductor part of a r.f. cavity. Measurements of the resonant frequency of the cavity are taken in a field $> B_c$ and in zero field. In the superconducting state

---

† The normal electron current also has an inductive component but this is extremely small.

‡ A derivation is given by Bleaney and Bleaney (1976), *Electricity and Magnetism* (3rd edn.), p. 236, Clarendon Press, Oxford.

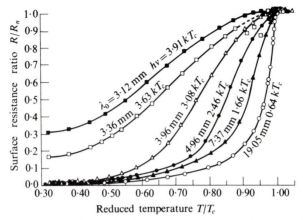

FIG. 14.13. The values of the ratio of the surface resistance in the superconducting state to that in the normal state for Al. The wavelengths and the corresponding photon energies are shown by each curve. Note that when the photon energy is $3.63\,kT_c$ or higher, the resistance *never* becomes zero. This is because the photons are able to excite the electron pairs across the energy gap (Biondi and Garfunkel, 1959).

the extra inductive reactance of the paired electrons changes the resonant frequency of the cavity. For temperatures well below T_c this frequency change, Δv, is of the form

$$\Delta v = \text{constant} \times (d_n - \lambda),$$

where d_n is the skin depth in the normal state. Since at low temperatures the normal resistance is constant (section 8.3), d_n is also constant. Thus the variation of Δv gives a direct indication of the changes in λ and this can be done to an accuracy of 0.1 per cent.

14.17. Coherence

Fundamental to our understanding of superconductivity is the concept of *coherence*—the idea that superconductivity is due to the mutual interaction and correlation of the behaviour of electrons which extends over a considerable distance. The existence of paired electrons whose states are correlated is central to the coherence hypothesis, but what we must now discuss is the *long-range* nature of this correlation; i.e. the paired electrons can be many thousands of atomic spacings (say 10^{-6} m) apart.

There are several observations and arguments (many suggested by Pippard before the development of the B.C.S. theory) which suggest that the superconducting state is one of long-range order. This order extends over a distance, the coherence length, which is denoted by ξ.† These are some of the reasons:

(a) The superconducting transition at T_c can be very sharp—within a range of 10^{-3} K or less. If superconductivity was due to electrons individually going into some new state, the statistical fluctuations would tend to give a broad transition. This is analogous to the sharp transition to the ferromagnetic state at the Curie temperature which is due to a cooperative interaction (section 12.4).

(b) The following argument based on the uncertainty principle may be used to deduce that the theory should contain a long-range parameter. It would seem reasonable that the electron states which are responsible for superconductivity should be those which lie within kT_c of the Fermi surface. Then those states must be defined to within kT_c and hence by the uncertainty principle their lifetime, τ, will be given by

$$(kT_c)\tau \approx \hbar$$

and hence if v is the electron velocity the wave function must extend over a distance $l = v\tau$. Therefore

$$l \approx \hbar v/(kT_c).$$

If we assume that v is of the order of the Fermi velocity ($\sim 10^6$ m s^{-1}) then l is about 10^{-6} m.

(c) The intermediate state of a type-I superconductor consists of a relatively small number of normal and superconducting regions (section 14.12). This suggests that the surface energy associated with the boundary is positive because this will minimize the surface area between the two states. We shall show in section 14.19 that for the surface energy to be positive, the boundary region between the normal and the superconducting sections must have an effective thickness, ξ, which is greater than the penetration depth;‡ i.e. the transition to a superconducting region is gradual because the density of pairs cannot vary rapidly and hence a finite thickness, of order ξ, is necessary to fully establish the superconducting state in the bulk material.

(d) The clearest indication of long-range order, however, is given by measurements which show that, contrary to the predictions of the London theory, the penetration depth depends on the impurity content of a particular sample and hence on the electron mean free path in the normal state (Fig. 14.14). However, the London relation $\lambda = (m/\mu_0 n_s e^2)^{\frac{1}{2}}$ does *not* involve the mean free path. Pippard suggested that the London equation (14.9) which relates the current density at a point to the field at that *same* point (i.e. it is a

---

† There are several different coherence lengths referred to in the literature, which depend on the theory or the phenomenon which is being discussed, but we shall not differentiate between them in our treatment.

‡ This is only true for type-I superconductors.

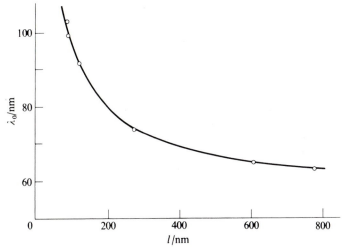

FIG. 14.14. The dependence of the low temperature penetration depth, λ_0, on the electron mean free path, l. This illustrates that the superconducting properties are dependent on l and a detailed interpretation of these results gave rise to the concept of coherence, or long-range order, between electrons (Pippard, 1953).

local relationship) should be replaced by an integral equation in which one can only determine the *average* value of j_s over a volume of order ξ^3—a coherence volume. Impurities will upset the electron coherence and they will therefore decrease the value of ξ. This modification to the theory was able to account for the experimental observations. It would therefore appear that the properties of a superconductor depend on the correlation of electrons within a volume ξ^3.

14.18. The Ginzburg–Landau theory

A completely fresh approach to the problem of calculating the equilibrium state of a superconductor when it is in an external magnetic field and the effect of field penetration was proposed by Ginzburg and Landau in 1950. In their theory the idea of long-range order is fundamental. A detailed discussion of this work would be out of place in this text, but since in any further reading on superconductivity reference would almost certainly be made to the theory, we give here the briefest outline.†

---

† Introductory treatments are given by Rose-Innes and Rhoderick (1978) section 8.5, and Solymar and Walsh (1984) p. 379. More detailed discussions are given by Kittel (1986) p. 633, Lynton (1969) chapter 5, Tilley and Tilley (1986) chapter 8, and Tinkham (1975) chapter 4. See Further Reading for full references.

The theory introduces an *order parameter*, ψ, whose modulus^2 is equal to the density of paired electrons. For a given T this is assumed to be a function of position in the material—it is not constant. The reason for this is that the theory has been developed in order to study the properties of surface layers and the boundary between normal and superconducting regions and this is just where one would expect an order parameter to vary. The equilibrium state of the superconductor in a field is determined by minimizing the free energy, G, for the whole sample. G is now not only a function of B (e.g. as in eqn 14.4) but also of ψ. Now in any physical problem, if parameters change rapidly extra energy is usually involved. The minimum G therefore contains a length, the coherence length again, over which ψ only varies slowly. A full development of the theory enables us to calculate the effect of a field on the penetration depth, λ, and the value of λ for very small samples. Its most remarkable achievement, however, was to predict the behaviour of type-II superconductors with their two critical fields, B_{c1} and B_{c2}, and the criterion which determines whether a material is type I or type II.

14.19. The criterion for type-I and type-II superconductivity

We have already described that observations of the intermediate state in type-I superconductors (which is resistive) show that a relatively small number of normal and superconducting regions are formed and this would suggest that the surface energy at the boundary between the two phases is *positive*. In type-II materials the penetration of a magnetic field is *not* accompanied by any resistance and the specimen breaks up into a fine filamentary structure of normal regions. It is believed that each of these filaments is at the centre of one flux quantum† and this induces a persistent current which circulates around the filament. This very fine filamentary structure would suggest that in type-II materials the boundary energy would be *negative* since the surface area appears to be maximized.

We now wish to show why the sign of the surface energy is controlled by the relative values of two dimensions which we have already discussed in some detail: the penetration depth, λ, and the coherence length, ξ. If ξ/λ is greater than unity‡ the surface energy will be positive and if it is less than unity it will be negative. To see how this arises, consider the boundary between a normal and a superconducting region in a field B_{ext}. The two phases are in thermodynamic equilibrium and so in the bulk their free energies must be equal. In the normal region the field is B_c. Within the boundaries of the superconducting phase there are circulating currents which oppose B_{ext}. These will increase the free energy, but this increase will be exactly compensated by the reduction in energy due to the electron pairing (Fig. 14.15(a)). In the boundary region, however, this compensation is not exact. The increase in

† See section 14.23.
‡ More strictly, according to the Ginzburg–Landau theory the ratio should be $\sqrt{2}$.

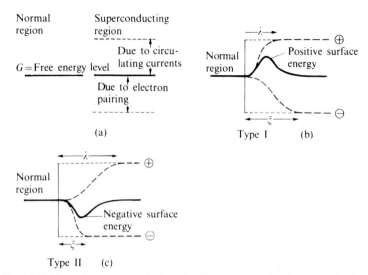

FIG. 14.15. The surface energy at the boundary between normal and superconducting regions. (a) The free energy in each phase is the same, but in the superconducting phase the increase in energy due to circulating currents is cancelled by the decrease due to electron pairing. (b) In type-I materials $\xi > \lambda$. Cancellation in the boundary region is not complete and there is a net increase in energy. (c) In type-II materials $\xi < \lambda$ and this leads to a net decrease in energy in the boundary region.

energy due to the circulating currents will occur over a distance of the order of their penetration depth, λ, whereas the decrease due to electron pairing will extend over the coherence length, ξ. Fig. 14.15(b) shows the situation if $\xi > \lambda$. There is an energy *maximum* within the interphase region. The case for $\xi < \lambda$ is shown in Fig. 14.15(c) and now there is an energy *minimum*, which is negative with respect to the bulk energy in the boundary layer.

Thus the relative magnitude of λ and ξ determines the sign of the surface energy and hence whether a material is type I or type II.

14.20. Flux motion and flux pinning

As the field applied to a type-II material is increased more normal filaments will be formed, but it might be thought, since there is a superconducting path through the material, that it would still have zero resistance. This, however, is not so because the electromagnetic forces due to the current act on the normal flux filaments so that they move through the material as if they were in a viscous medium, and, in a manner analogous to the irreversible motion of domain walls (section 12.10), this motion dissipates energy. A potential appears across the specimen and some of its resistance returns.

This problem may be overcome by *flux pinning*. It is achieved by introducing a large number of defects into the crystal, usually by severe cold work. Even so, the number of materials which can be fabricated into wire or strip and which can carry large currents in very high fields is not large. The best material that is readily available is Nb_3Sn, which has a B_{c2} of about 20 T and which can sustain a current density of about 10^5 A cm^{-2} in about 10 T. A cheaper alternative which is satisfactory up to about 9 T is a Nb–Ti (roughly 1 : 1) alloy.

14.21. Energy gap measurements; infrared and tunnelling experiments

We have already described how the energy gap, 2Δ, between the paired and the unpaired electron states gives rise to an exponential temperature dependence for the electronic specific heat in the superconducting state at low temperatures. We now discuss two other types of experiment which both demonstrate the existence of the energy gap and enable it to be measured directly.

The first is to investigate the response of a superconductor to infrared radiation as the photon energy is varied in the region of 2Δ. A photon with this energy will be able to break up the electron pairs. We have already given one example of this in Fig. 14.13 where it is shown that the surface resistance in the superconducting state never drops to zero if the photon energy is $3.63\, kT_c$ or greater. A more direct technique is to measure the infrared reflectivity† of a superconductor. It is found that this drops sharply to the normal state value for photon energies greater than about $3.5\, kT_c$. This is the value of 2Δ which is predicted by the B.C.S. theory as T tends to 0 K. Since kT_c usually corresponds to a wavelength of about 0·5 mm the response of a superconductor to any shorter wavelength radiation, e.g. visible light, is exactly the same in both the normal and the superconducting states.

The other technique which enables the energy gap to be measured directly is one which detects quantum mechanical tunnelling (see the footnote on p. 119) of unpaired electrons through a very thin insulating layer (about 2 nm thick) which separates two superconducting films. In Fig. 14.16(a) the energy states for the electron pairs are shown as all being at one level at the lower edge of the energy gap and the excited single electron states are shown as a band starting at an energy Δ above the pair states.‡ At any non-zero temperature there will be a few occupied states in the single electron band.

---

† Transmission experiments have also been made, but since they must be done on films which are thinner than the penetration depth, the interpretation is more complicated.

‡ There is a slight problem in drawing this diagram because in this experiment we are interested in observing the tunnelling of *single* electrons. It is therefore necessary that all energies are shown as single electron energies. For this reason the energy gap is indicated as Δ and not as 2Δ.

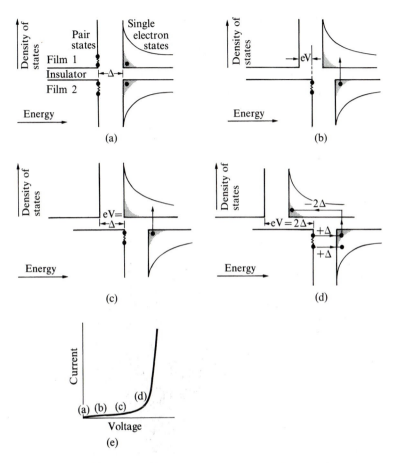

FIG. 14.16. Tunnelling between identical superconductors. (a) The potential, V, across the junction is zero. (b) For low V, the small number of single electrons present can tunnel from 2 to 1. (c) When the pair level of 2 is opposite the edge of the single electron band of 1 pair tunnelling cannot occur because energy is needed to separate the pair into single electrons. (d) When the relative energy displacement of the two films is 2Δ, the paired electrons can split up and tunnel with conservation of energy. (e) The voltage-current characteristic of the device. Note the increase in tunnelling current corresponding to (d).

The experiment involves the measurement of the tunnelling current from one film to the other as the potential between them is increased.† The general principle is similar to that of the tunnel diode (section 10.11). As in any tunnelling process the total energy must be conserved.

We now follow the sequence of events as the potential of film 1 is made positive with respect to film 2, so that all its energy states move to the left. At (b) only a small current can flow from 2 to 1 because the paired electrons on the left of 2 face the energy gap of 1 and only the few single electrons in the right-band of 2 can tunnel to vacant states in 1. In (c), where the relative shift in energy is Δ, the pair energy on the left of 2 is level with the edge of the right-band of 1 and it might be thought that the electron pairs could now tunnel from 2 to 1. However, since these electrons must tunnel into single electron states, each pair must first be split up at a cost of Δ per electron. Tunnelling does not occur, therefore, until (d), when the relative energy shift is 2Δ. This enables the energy of a pair to be conserved because, although the single electron which has *not* tunnelled has gained Δ, the one which has tunnelled loses Δ when it drops into the lowest occupied state of 1. The tunnelling current rises rapidly (e) and if at this point the potential between the films is V_g then eV_g is a measure of the energy gap 2Δ. This technique is the most direct method of measuring the gap and it is a convenient method of determining the variation of 2Δ with temperature (see Fig. 14.8).

14.22. Phase coherence

We have already discussed the coherence of electron pairs over reasonably large distances—of the order of 10^{-6} m in some cases. This is the coherence between the individual electrons which form the pairs. When, however, we consider the flow of a superconducting current a quite different, additional form of coherence arises. This occurs because the electron pairs cannot be scattered.

In a normal metal the free electron wave function can be described as a travelling wave of the form $\psi = A \exp(i\mathbf{k} \cdot \mathbf{r})$ (see the footnote on p. 108). The quantity in the exponent, $\mathbf{k} \cdot \mathbf{r}$, is the *phase* of the wave. Every time a normal electron is scattered its wave vector $\mathbf{k}$ will alter, and so, as the electron travels through the metal, its wave function will undergo many *random* changes of phase. Our knowledge of the phase of the wave function at one point does not help us to predict the phase at any other point. A superconducting system can also be considered to have a wave function, the Ginzburg–Landau order parameter (section 14.18), which contains a similar phase term, except that now $\mathbf{k}$ is the effective combined wave vector of the two electrons. However, because the pair cannot be scattered, the phase difference between points $\mathbf{r}_1$ and $\mathbf{r}_2$ will be $\mathbf{k} \cdot (\mathbf{r}_2 - \mathbf{r}_1)$ no matter how great the distance $\mathbf{r}_2 - \mathbf{r}_1$. This is called *phase coherence* and it has some profound consequences.

†We describe here a tunnelling experiment between the same two superconductors. Measurements have also been made between two different superconductors and between a superconductor and a normal metal, see problem 14.8.

14.23. Flux quantization

Consider a superconducting ring of inner radius R which has some magnetic flux, Φ, trapped inside it (Fig. 14.17). This flux as we have seen is produced by persistent currents which will be circulating on the inner surface of the ring.

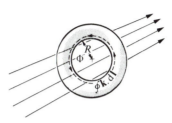

FIG. 14.17. Flux quantization. The phase coherence of a persistent current demands $\oint k \,.\, dl = 2\pi \times \text{integer}$ (eqn (14.21)). This leads to the requirement that the flux, Φ, in the loop is quantized in units of $(h/2e)$.

As discussed in the previous section, there must be phase coherence at any point on the inner circumference no matter how many times the pairs circulate, and at any point the phase must remain constant. Hence the integral of the phase of the order parameter around the inner circumference must be an integer $\times 2\pi$; i.e.

$$\oint_R k \,.\, dl = 2\pi n, \qquad (14.21)$$

where n is an integer.[†] We now need to relate k to the superconducting current density, j_s, and to the flux.

We first recall (section 7.3) that the momentum of a free electron is related to k by $\hbar k = mv$. In the presence of a magnetic field, however, this relationship is modified[‡] to $\hbar k = mv + eA$, where A is the vector potential of the field B and is defined by $B = \text{curl } A$. For an electron pair this becomes

$$\hbar k = 2mv + 2eA.$$

We write j_s in the form

$$j_s = n_s ev,$$

where n_s is the density of paired electrons, i.e. twice the number of pairs. Then k may be expressed as

$$k = (1/\hbar)(2mj_s/n_s e) + (2e/\hbar)A.$$

---

† This is analogous to the calculation of stationary orbits in the Bohr theory of the atom.

‡ See Bleaney and Bleaney (1976), *Electricity and Magnetism* (3rd edn.), section 5.1, Clarendon Press, Oxford. e is here the *modulus* of the electronic charge.

We now integrate around the circumference at R, as in eqn (14.21) and obtain

$$\oint_R \boldsymbol{k} \cdot \mathrm{d}\boldsymbol{l} = 2m/(n_s e\hbar) \oint_R \boldsymbol{j}_s \cdot \mathrm{d}\boldsymbol{l} + (2e/\hbar) \oint_R \boldsymbol{A} \cdot \mathrm{d}\boldsymbol{l} \doteq 2\pi n. \tag{14.22}$$

Since $\boldsymbol{B} = \operatorname{curl}\boldsymbol{A}$ we may use Stokes' theorem† to change the line integral of $\boldsymbol{A}$ into an area integral of $\boldsymbol{B}$, and so

$$2m/(n_s e\hbar) \oint_R \boldsymbol{j}_s \cdot \mathrm{d}\boldsymbol{l} + (2e/\hbar) \int \boldsymbol{B} \cdot \mathrm{d}\boldsymbol{S} = 2\pi n, \tag{14.23}$$

or

$$m/(n_s e^2) \oint_R \boldsymbol{j}_s \cdot \mathrm{d}\boldsymbol{l} + \int \boldsymbol{B} \cdot \mathrm{d}\boldsymbol{S} = n(h/2e). \tag{14.24}$$

But the integral $\int \boldsymbol{B} \cdot \mathrm{d}\boldsymbol{S}$ is just the flux, Φ, within the inner circumference. If we take the line integral of $\boldsymbol{j}_s$ at a radius slightly larger than R, well outside the penetration depth, then $\boldsymbol{j}_s$ will be zero and we achieve the final simple result that

$$\Phi = n(h/2e) = n\Phi_0. \tag{14.25}$$

Thus the flux within the superconducting loop is *quantized* in units of $\Phi_0 = h/2e$. If we do not neglect the line integral of $\boldsymbol{j}_s$ in eqn (14.24), the sum of the two terms on the left hand side is sometimes called the *fluxoid* and strictly speaking, it is the fluxoid which is quantized.

The individual flux quantum is very small, $2 \cdot 07 \times 10^{-15}$ weber. It has been detected by sensitive magnetic measurements and its value demonstrates conclusively that the superconducting current carriers are *pairs* of electrons and not single ones.

14.24. Josephson effects and quantum interference

In section 14.21 we described the process whereby *single* electrons could tunnel from one superconducting film to another via an insulating layer. Josephson predicted that if this layer was sufficiently thin, superconducting *pairs* could tunnel through it with zero potential difference across the junction. There would be phase coherence from one side to the other, although there would be a phase change $\Delta\phi$ across the junction. He showed that the tunnelling current would have the form

$$I = I_0 \sin(\Delta\phi), \tag{14.26}$$

where I_0 is the maximum current that the junction can carry without a potential difference across it. If the current is controlled by an external circuit the

† $\oint \boldsymbol{A} \cdot \mathrm{d}\boldsymbol{l} = \int \operatorname{curl}\boldsymbol{A} \cdot \mathrm{d}\boldsymbol{S} = \int \boldsymbol{B} \cdot \mathrm{d}\boldsymbol{S}$.

electron pairs will adjust their phase, $\Delta\phi$, so as just to accommodate the current. I_0 will depend on the properties of the particular barrier, e.g. its thickness. In practice the barrier does not have to be an insulating film. It is often a very small adjustable screw contact and it is sometimes called a *weak link*. The properties of these superconducting weak links or Josephson junctions have been exploited in several devices which are basically used to measure very small changes in magnetic flux in units of the flux quantum. We shall describe a relatively simple one which demonstrates the general principle.

Consider a superconducting loop which has connexions to an external current supply via two diametrically opposed leads and which contains *two* weak links X and Y (Fig. 14.18(a)). Let a current, I, which is less than $2I_0$, flow

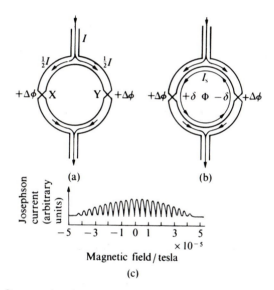

FIG. 14.18. Quantum interference. (a) A current-carrying superconducting circuit containing weak links at X and Y has a phase change, $\Delta\phi$, at each link. (b) If a flux Φ is introduced into the circuit the circulating persistent current it induces will give an extra phase change $+\delta$ at X and $-\delta$ at Y. (c) The current-carrying characteristic of such a device as a function of Φ shows a variation which has a period of $(h/2e)$ (Jaklevic, Lambe, Mercereau, and Silver, 1965).

from one lead to the other through the loop and, for simplicity, let us assume that the two weak links are identical so that $\frac{1}{2}I$ flows through each with a phase change $\Delta\phi$. We now introduce some magnetic flux into the loop, e.g. by energizing a small coil. This will induce a persistent current, I_s, round the loop which will *add* to the current and the phase change at X, but which will *subtract* from them at Y (Fig. 14.18(b)). The two currents will, from eqn (14.26), be given by

$$\tfrac{1}{2}I + I_s = I_0 \sin(\Delta\phi + \delta) \quad \text{and} \quad \tfrac{1}{2}I - I_s = I_0 \sin(\Delta\phi - \delta),$$

where δ is the extra phase change introduced by the persistent current. The total current through the device will still be I, but it will be defined by

$$I = I_0 \{\sin(\Delta\phi + \delta) + \sin(\Delta\phi - \delta)\},$$

which by straightforward manipulation may be written as

$$I = 2I_0 \cos\delta \sin\Delta\phi. \tag{14.27}$$

We now need to calculate the extra phase change, δ, in terms of the magnetic flux within the loop. This we can do from eqn (14.23) since the term involving the line integral of j_s is just the phase change due to the superconducting current. If we assume that this integral is taken over a path well inside the superconductor then, except at the links themselves, j_s will be zero and so

$$2\pi n - (2e/\hbar) \int \boldsymbol{B}.\,\mathrm{d}\boldsymbol{S} = \text{total phase change at both links} = 2\delta. \tag{14.28}$$

Since $\int \boldsymbol{B}.\,\mathrm{d}\boldsymbol{S} = \Phi$, we then have, using (14.25)

$$\delta = \pi\{n - (2e/\hbar)\Phi\} = \pi(n - \Phi/\Phi_0).$$

If, initially, the flux was zero and we increase it very slowly, we may first put $n = 0$ and we then have for the current through the device

$$I = 2I_0 \cos(\pi\Phi/\Phi_0) \sin\Delta\phi. \tag{14.29}$$

On comparing this expression with eqn (14.27) we see that as the flux is increased, the maximum current through the device varies periodically with Φ. There is just one slight complication. As eqn (14.29) stands it would appear that as the cosine term becomes negative the total current through the device would reverse. This cannot occur if it is being driven externally. The situation is actually saved by the electron pairs themselves, which will adjust their phase, $\Delta\phi$, to ensure that the current flows in the correct direction. We may take account of this most simply by using the modulus of the cosine term in eqn (14.29)

$$I = 2I_0 |\cos(\pi\Phi/\Phi_0)| \sin\Delta\phi.$$

Thus the maximum current varies periodically with the flux through it with a period which is just one flux quantum (Fig. 14.18(c)). It will be recognized that this is an analogue of a two-slit optical interferometer and hence these devices are sometimes called quantum interferometers.

There are several other devices in which weak links can be used to detect flux quanta.† Some are very sophisticated and are fiendishly difficult to understand. The one described here is by far the simplest.

14.25. The a.c. Josephson effect

The tunnelling current defined by eqn (14.26) cannot exceed I_0 unless a potential $-V$ is applied across the junction. An electron pair will then gain energy $2eV$ when it tunnels through. This energy cannot be dissipated by collisions but a *photon* of energy $h\omega = 2eV$ can be radiated. These photons can be detected and since

$$\omega = (2e/h)V,$$

a measurement of the frequency as a function of V enables us to obtain a value for the ratio $2e/h$. Since if $V = 1 \mu V$ the frequency is 484 MHz, which lies in a convenient range, it is possible to obtain a far more accurate value (to about 1 part in 10^6) for the ratio of these two fundamental constants than can be achieved by any other method.

The converse effect also occurs. Radiation (especially in the infrared or microwave region) can be absorbed by the junction and this will excite an extra potential across it.

14.26 Amorphous superconductors

The first amorphous material to be shown to be superconducting was Bi (Bückel and Hilsch 1954). This, when condensed from the vapour onto a substrate that was cooled to liquid helium temperatures, was shown to be a superconductor with a transition temperature, T_c, of about 6 K. However, on warming up it crystallized, became a semi-metal, and the superconducting property was lost. Since then many amorphous materials (mainly alloys) have been shown to be superconductors with values of T_c ranging up to about 9 K. In a magnetic field all of them exhibit type 2 characteristics (section 14.6, p. 237; i.e. the flux starts to penetrate at a lower critical field, but the electrical resistance does not return until a higher critical field is attained.

So far none of these materials appear to be useful in technology because, due to their homogeneity, they do not have the capability of adequate flux-pinning (section 14.20) and so the passage of an electric current easily destroys their superconductivity.

The reason why these materials are all type 2 is quite simple. We have already seen (section 14.19, p. 256) that the criterion for type 2 behaviour is

† They have such exotic names as SQUIDS and SLUGS.

that the wave function coherence length, ξ, should be shorter than the magnetic penetration depth, λ. In the amorphous state wave function coherence will only operate over a short distance and hence the condition $\xi < \lambda$ will commonly occur.

14.27 Summary

This has been a complicated chapter. There are so many facets to superconductivity that it is worth recollecting the most important points.

Superconductivity, the disappearance of electrical resistivity at T_c, is accompanied by the expulsion of any magnetic flux from within the specimen —the Meissner effect. Superconductivity can be destroyed in a field greater than B_c and this tends to zero as T_c is approached. Well within the specimen $B_{int} = 0$. In the presence of an external field this is achieved by persistent surface currents which flow in a layer of finite thickness, λ, and hence the field does penetrate a distance λ.

The superconducting state is characterised by a pairing of electrons due to an interaction with the ionic lattice. The number of paired electrons tends to zero as T approaches T_c.

There is an energy gap 2Δ between the paired electron states and the single-electron states which tends to zero as T_c is approached.

The distance between an electron pair can be quite large—the coherence length ξ.

If $\xi/\lambda > 1$, there is a positive surface energy between normal and superconducting regions in an external field and we have type-I superconductivity. This is characterised by a relatively sharp transition to the normal state at B_c. If $\xi/\lambda < 1$ the energy is negative and the material is type II. It has two critical fields, B_{c1} and B_{c2} between which a mixed state develops. This is formed of normal filaments in a superconducting matrix. The resistance remains zero up to B_{c2}.

A superconducting current exhibits long-range phase coherence. In a closed loop this gives rise to flux quantization and this can be used in various devices to detect flux quanta.

PROBLEMS

14.1 From the data in Fig. 14.2 estimate, for $1 \, \text{cm}^3$ of tin, (a) the difference in the free energy between the normal and the superconducting states at $0 \, \text{K}$, and (b) the specific heat discontinuity at the superconducting transition temperature.

14.2. Show that when superconductivity is destroyed by the application of a magnetic field the material will cool.

A sample of tin is thermally isolated at $2 \, \text{K}$ and a field $> B_c$ is applied. Calculate the final temperature of the specimen (neglect the lattice specific heat). [For tin: density $= 7300 \, \text{kg m}^{-3}$; electronic specific heat $= \gamma T$, where $\gamma = 1.75 \times 10^{-3} \, \text{J g mol}^{-1} \, \text{K}^{-1}$, atomic weight $= 119$.]

14.3. Derive the second London equation, $\partial j_s/\partial t = \mathscr{E}/\Lambda$ (14.10) and show that $\Lambda = m/(n_s e^2)$. (Hint: use Newton's second law and assume that $j_s = n_s ev$). N.B. This derivation is not rigorous, see Tinkham (1975) *Introduction to superconductivity*, section 1.2.

14.4. Estimate the ratio of the thermal conductivity in the normal state to that in the superconducting state at 10^{-2} K for an annealed lead strip $0\cdot1$ mm thick, if its residual electrical resistivity in the normal state is 2×10^{-9} ohm m. (See sections 6, 10 and 8.6). [For lead: velocity of sound $= 2000$ m s^{-1}; density $= 11300$ kg m^{-3}; Debye $\theta \approx 108$ K.]

14.5. The penetration depth, λ, of Hg at $3\cdot5$ K is about 75 nm. Estimate the values, as T tends to 0 K, of (a) λ, and (b) n_s.

14.6. Show that, due to field penetration, the critical field of a superconducting slab of thickness t is of the order of $B_c(1 + \lambda/t)$, where B_c is the critical field of a bulk sample. (Hint: assume that the free energy difference between the normal and the superconducting states is independent of the specimen size.)

14.7. A superconducting solenoid, cooled in liquid helium, has 20 000 turns and is 15 cm long. Its mean diameter is 3 cm. The current through the solenoid is increased slowly and when it reaches 40 A the solenoid goes normal. Estimate the volume of liquid helium which is evaporated. [Latent heat of liquid helium $= 2\cdot5$ J cm^{-3}.]

14.8. A tunnel junction, similar to that described in section 14.21, is made from two *different* superconducting films which have energy gaps $2\Delta_1$ and $2\Delta_2(\Delta_1 < \Delta_2)$. They are cooled until they are both superconducting to a temperature which is just slightly below T_c for film 1. The potential between the films is increased with film 1 conventionally negative with respect to film 2. Show that the current–voltage characteristic of the junction has a local maximum when $V = (\Delta_2 - \Delta_1)/e$ followed by a local minimum at $V = (\Delta_1 + \Delta_2)/e$. (See I. Giaver, *Phys. Rev. Letters* **5**, 464 (1960).)

14.9. Each black spot in Fig. 14.5 indicates the position of a filament of normal material and it contains one flux quantum. Estimate the average field on the specimen.

Assume that the average field within each filament is of the order of B_{c2} $(0\cdot2$ T)—this is only a rough assumption—and hence calculate the effective diameter of each filament. (Note: this should be about a coherence length). Hence estimate the volume magnetic susceptibility (m_v/H) of the specimen when it was in the state indicated in the figure.

14.10. A thin superconducting ring of inner diameter 1 cm has 10 flux quanta trapped in its central hole. If the penetration depth is 50 nm estimate the value of the persistent current which circulates in the ring. [The self-inductance of a loop of radius R made of wire of cross-sectional radius r is, for $R \gg r$, equal to $\mu_0 R \log_e (8R/r)$ henry.]

14.11. The electrical resistivity of a Nb cable when in the normal state is 3×10^{-9} ohm m at 8 K. Estimate the power dissipation in 1 km of superconducting Nb cable, 10 mm in diameter, maintained at 8 K, when it is carrying a current of 1000 A at 50 Hz. Assume that the cable is in the fully superconducting (i.e. not mixed) state. [For Nb: $T_c = 9\cdot3$ K, $\lambda_0 = 47$ nm]. (Note that the very small energy loss as calculated in this problem would hardly be serious. A much larger dissipation occurs in practice and is due to the oscillatory viscous motion of the normal filaments in the mixed state, see section 14.20.)

15. Amorphous materials

15.1 Introduction

Amorphous materials such as glasses and many polymers exhibit no complete regularity in the arrangement of their atoms or molecules and this is the way in which they differ from crystalline solids. In a crystal, as we have seen in Chapter 1, the atoms are arranged as a regular network on a lattice and so if we know the pattern of atoms in one region of the material then we can predict precisely where the atoms at some other part of the crystal should be. Occasionally where the atoms are not at these predicted points we say that the crystal contains *defects* (Chapters 3 and 4).

In amorphous solids the position of the atoms does not have such regularity but this does not mean that they are packed together in a random mess. The interatomic forces and the bonds between atoms will still be very similar to those in crystals and this will ensure that the atomic separations and the number and arrangement of nearest neighbours of an atom will on average be the same at any particular point of the sample so that the environment of one atom will be very similar, but not necessarily exactly the same, as that of another. On account of these very slight changes in spacing and orientation throughout the material it follows that the precise positions of distant atoms cannot be predicted.

Of course it is not possible to define a lattice to describe such a structure and so we need to modify our ideas on all those properties that we have described in earlier chapters that are dependent on the material having a lattice. To a large degree we shall be discussing the differences and similarities between amorphous and crystalline materials; to avoid repetition considerable reference will be made to earlier chapters. At the outset it should be stated that whilst our main concern is to describe and (wherever possible) to explain some of the most important properties of amorphous materials, the intense current interest in this field is in the search for *new materials* which either have novel properties or are cheaper to produce in bulk than their crystalline counterparts.

15.2 Types of amorphous materials

Amorphous materials can be divided into two main groups. Firstly there are those which, when they are made, tend to be produced in the amorphous state. These are the natural glass-forming materials. There are, however, other materials that would normally solidify as a crystalline structure, but that can be made in an amorphous form, either by rapidly quenching the molten material or by allowing its vapour to condense onto a cold surface. The amorphous nature of the natural glass-formers is quite stable (although prolonged heating can enable some of the simpler materials to crystallize), but the materials that can only be made by rapid cooling of the melt or vapour usually crystallize quite rapidly when they are heated above a certain temperature.

One of the simplest and most common of the natural glass-forming materials is amorphous silica, SiO_2, which also exists in crystalline forms as crystobalite and quartz. Ordinary window glass (soda glass) is mainly SiO_2, plus Na_2O and CaO, but there are also thousands of formulations of glasses for special purposes which have additions of other materials to change the optical, electrical, mechanical and thermal properties. Other glasses which are of particular interest because they are semiconductors, are compounds of S, Se, or Te with elements such as As and Ge. These are called chalcogenide glasses.[†]

Another group of natural glass-forming materials are those that consist of very large molecules (macromolecules). Such molecules cannot rotate and fit together easily (this is called steric hindrance) and so they cannot form crystals. Simple examples are glycerol and glucose, but many polymers such as polystyrene, rubbers, and epoxy resins are amorphous.

15.3 The preparation of amorphous materials

The main amorphous materials produced by rapid cooling are pure metals, alloys, and semiconducting elements and compounds. There are several different methods of manufacture and the conditions for satisfactory production tend to be different for each material. One very successful technique is to spray a jet of molten material on to the rim of a rapidly rotating copper wheel or between two rollers, where it solidifies. This solid is continually pulled away and hence a long ribbon of amorphous material is formed. This is called *splat cooling*. Small samples can also be made by quenching the melt in water and also by rapid compression of the melt by a hammer and anvil device. Other techniques involve the deposition of the vapour onto a cold surface; this can be achieved by chemical deposition, by cathodic sputtering or by the breakup of a molecule by electrical discharge. For example, silicon is produced by the r.f. glow discharge of the gas silane, SiH_4, and this can be deposited as amorphous silicon on a suitable surface. The object of all these techniques is to 'freeze' the atoms in random positions at a temperature so low that the thermal energy is not sufficient for them to rearrange themselves into a crystalline form. In fact amorphous films of many metals can be produced by condensing the vapour onto a surface that has been cooled to the temperature of liquid nitrogen or liquid helium, but these nearly always crystallize when they are warmed to room temperature.

† Strictly speaking the chalcogenides are compounds of any element of group 6 of the periodic table, but the term chalcogenide glass does not usually include oxide glasses.

15.4 Devitrification

The crystalline arrangement of atoms has the lowest free energy, so all amorphous materials are in a metastable state. Hence there will always be a tendency, even with natural glass-formers, for crystallization to occur, provided that sufficient thermal energy is available to unlock the atoms to enable them to move to other sites. This is called *devitrification*. At room temperature it is usually a very slow process, although ordinary soda glass does partially devitrify over a period of many years and the process can be accelerated if the glass is held for some time at a high temperature.

Materials that need to be formed by rapid quenching usually crystallize very rapidly above a certain temperature. Heat is evolved during this transition, which can be detected by an anomalous rise in the temperature of the sample whilst crystallization is occurring (Fig. 15.1).

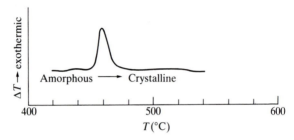

FIG 15.1. The evolution of heat in $Fe_{79}SnB_{20}$ as it reverts to the crystalline state on being heated through 460°C. (Dunlap and Stroink, *J.Appl.Phys.* **55**, 1068 (1984).

15.5 Structural models

The idea that an amorphous solid could have a definite type of *structure* might at first appear rather strange. Nevertheless if we wish to develop detailed theories it is very useful to have models of possible atomic arrangements. These should take account of the fact, already discussed, that the nearest neighbour atomic arrangement must be similar to that of the crystalline material. Such models can be divided into two groups: (a) those for materials in which the bonding between atoms is *not directional*, as for example in ionic materials and metals; and (b) those materials that have *directional* bonding, as in covalently bonded materials such as Si or SiO_2.

For non-directional bonding the simplest model is the *dense random packing of spheres* (Fig. 15.2(a)). This consists of an assembly of identical spheres, so constructed that when another sphere is added, it has to touch three neighbouring spheres in such a way that no voids are left that could accommodate another sphere. Such models have been made with hundreds or even thousands of ball-bearings, their coordinates were recorded and their

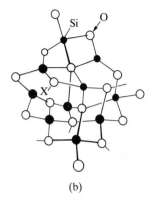

(a)   (b)

FIG 15.2. Simple structural models for amorphous materials. (a) Dense random packing of spheres, for materials that do not have directional bonding. (b) Continuous random network for directionally bonded materials, e.g. SiO_2; this is based on tetrahedral units with some variations in bond length and bond angle. X indicates a *dangling bond* (section 15.6).

diffraction patterns (section 15.11) calculated. This painstaking work has now been replaced by computer simulations of these structures.

For directional bonding the simplest model is the continuous random network. As an example let us take SiO_2, which is based on an assembly of tetrahedral units of SiO_4. These are connected together, not as in the crystalline arrangement of Fig. 1.6(b), p. 6, but instead with small random variations of angle between one tetrahedron and its neighbour (Fig.15.2(b)). This involves a certain amount of bond bending and it is also necessary to slightly vary the bond lengths.

These models are some of the simplest that can be derived and they certainly do not cover all types of amorphous materials. In particular they do not simulate the structure of many organic polymers or other materials composed of very large, complex molecules. These do not exist in any crystalline form that could be used as the basis for constructing an amorphous model.

15.6 Defects, dangling bonds

Since an amorphous structure has randomness as its main feature, the possibility that it could contain defects might appear surprising. But of course the simple models described in the previous section are idealized. There might be an atom missing here or there and this would create a vacancy. In the continuous random network a bond might not find a partner. It is then broken and it is called a *dangling bond* (Fig. 15.2(b)). This bond then lacks the second electron which would complete the electron shell and so it is liable to trap any impurity with an electron that could satisfy this need.

The presence of these and other defects is thought to be very important in influencing the behaviour of amorphous semiconductors (section 15.18 *et seq.*).

15.7 Homogeneity

How widely-spread is the disorder in the amorphous state? Does it have any regularity at all or does it exhibit complete randomness down to the molecular level? Over large distances there is no doubt that the material is disordered, but under high magnification in the electron microscope, structure can be observed in some materials. It is still not completely clear whether clusters, such as those shown in Fig. 15.3, which are of the size of some tens to a hundred nm, themselves contain any ordered material or whether they are completely amorphous. In some cases X-ray or neutron diffraction studies also suggest the presence of regions of slightly different structure, density or of tiny voids, but nevertheless on the whole most amorphous materials are very homogeneous and defect-free.

Certain polymers, e.g. polyethylene, can have some regions which are definitely crystalline and others which are amorphous. The degree of crystallinity can be increased by extruding the material through a die.

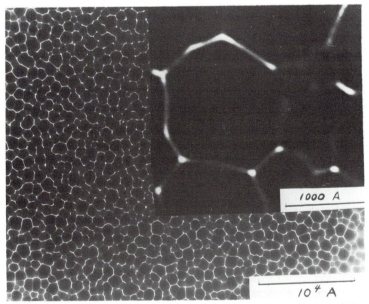

FIG 15.3. Electron micrograph of clusters in amorphous As_2Se_3 film. (Chen. Phillips, Bridenhaugh, and Aboav, *J. Non-cryst. Solids* **65**, 1 (1984).)

15.8 Mechanical properties

Amorphous materials exhibit a wide variety of mechanical behaviour—ranging from the extreme elasticity of rubber and the plastic deformation of polystyrene to the brittle fracture of glass. Nevertheless, in spite of these differences, all share a feature in common with crystals—the stress for fracture is far less than the theoretical shear stress of the material (section 3.10, p. 45).

At a sufficiently low temperature (room temperature for many materials) all amorphous materials become brittle, as is typified by glass. This is because the energy which would be required to move atoms or groups of atoms past one another in any type of slip mechanism that would give rise to plasticity is then much greater than the energy which is required to make cracks run through the material. Hence in these circumstances when stress is applied, the cracks extend before any plastic flow can occur and brittle failure follows. This process, which has been briefly discussed in section 4.23, p. 68, is now analysed in more detail.

15.9 The Griffith theory of crack propagation

Griffith proposed that brittle fracture is initiated by the growth of surface cracks which, under the influence of an applied force, run rapidly through the material. He suggested that the criterion for the stress to make the crack extend (in preference to elastic or plastic deformation) is that the surface energy of the additional area of the crack faces which would then be formed, must be equal to the elastic strain energy which was stored in that region before the crack had spread.

If a surface crack is present, it acts as a *stress concentrator* so that the stress in the material near the tip of the crack is much greater than the applied stress. Let us consider a crack normal to the length of a long rectangular bar of unit thickness which is subject to a tensile stress, σ (Fig. 15.4). If the crack has a depth, L, and the radius of curvature at its tip is r then the stress in the region of the tip is magnified by a factor of $\sqrt{(L/r)}$. This stress concentration factor can easily have a value of over 1000.

A crack has a surface energy that is analogous to the surface energy (surface tension) of a liquid. We denote it by γ. If this crack (of unit length) grows into the material by ΔL the additional surface energy of the two faces so formed will be

$$(\Delta E)_{\text{surface}} = 2\gamma \, \Delta L. \tag{15.1}$$

The elastic energy of a body subjected to a uniaxial tensile stress, σ is $(1/2)\sigma^2/Y$ per unit volume, where Y is Young's modulus. When the crack extends, σ is relaxed over a certain volume and hence some elastic energy is released as surface energy. If this extension is ΔL, then for an order of magnitude

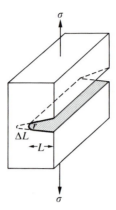

FIG 15.4. The Griffith theory of crack propagation. Under the influence of the stress, σ, a crack of depth L extends an extra ΔL into the material—see text.

calculation let us assume that the elastic energy in a cylindrical region of radius ΔL is released. This energy will be

$$(\Delta E)_{\text{elastic}} = (1/2)\sigma_t^2 \pi (\Delta L)^2 / Y \tag{15.2}$$

where σ_t is the stress at the tip of the crack. For the Griffith criterion to hold we need $(\Delta E)_{\text{surface}} = (\Delta E)_{\text{elastic}}$

i.e.

$$2\gamma \, \Delta L = (1/2)\sigma_t^2 \pi (\Delta L)^2 / Y \tag{15.3}$$

and so

$$\sigma_t^2 = 4\gamma Y / (\pi \, \Delta L). \tag{15.4}$$

To write this in terms of the applied stress, σ, we multiply by r/L, since the stress concentration factor is $\sqrt{(L/r)}$ and we then have

$$\sigma^2 = 4\gamma Yr / (\pi L \, \Delta L). \tag{15.5}$$

If we assume that the increased stress at the crack tip extends into the material to a distance of order r, then $r \approx \Delta L$ and we have as a condition for crack propagation that

$$\sigma^2 \approx \gamma Y / L. \tag{15.6}$$

Thus the fracture stress will be reduced for materials with deep cracks. This prediction is borne out by measurements on very fine fibres which, due to their small surface area, will have very few deep surface cracks. Such fibres have been found to have fracture strengths far greater than that of the bulk material. The effect of reducing the depth and number of cracks can also be demonstrated on glass by flame polishing or by treating the surface of a glass rod with very dilute hydrofluoric acid, which will also reduce the number and size of the cracks. After such procedures the glass can be bent into a much tighter radius of curvature before it fractures than when it is untreated.

15.10 The brittle–plastic transition

If, however, the glass is heated beyond a certain temperature it softens and can then be deformed plastically. This occurs because the thermal energy is sufficient to enable the atoms to move past one another, and so the stress required to make cracks run is not reached. The temperature at which this brittle–plastic transition occurs is called the glass transition temperature and it is usually denoted by T_g.

The fact that above T_g the atoms can move much more freely with respect to one another shows up clearly in the thermal expansion of glass. Above T_g the rate of expansion with temperature increases considerably (Fig. 15.5) and in fact the temperature at which this increase occurs is often used as a measure of T_g.

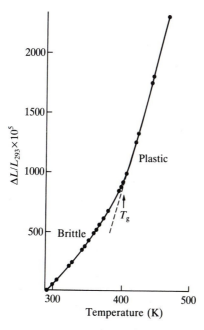

Fig 15.5. The brittle–plastic transition. At the transition temperature, T_g, there is a sharp change in the thermal expansion. The data are for epoxy-resin (J. Grubb, D.Phil thesis, Oxford (1975).)

15.11 Diffraction effects in amorphous materials

In order to study the results of X-ray, electron, or neutron diffraction experiments, let us first recall the general expression that we derived in section 2.5 (p. 21) for the total amplitude, at a point $\mathbf{r}$, of waves which are diffracted by an array of atoms at positions $\mathbf{r}'$

$$\text{Amplitude} = A' \exp(i\mathbf{k}.\mathbf{r}) \Sigma_{\mathbf{r}'} \exp\{i(\mathbf{k}_0 - \mathbf{k}).\mathbf{r}'\} \tag{15.7}$$

where $\mathbf{k}_0$ and $\mathbf{k}$ are respectively the incident and the diffracted wave vectors and A' is the amplitude of the scattered wave from a single atom. For a crystal the periodicity of the lattice and hence the periodicity of the values of $\mathbf{r}'$ enabled us to deduce the conditions for which (15.7) gives a non-zero value, but in an amorphous material there is no long-range regularity in the atomic positions. The mathematical treatment of this problem is more difficult and it requires a new type of averaging procedure.

In order to continue with the analysis we first multiply (15.7) by its complex conjugate. This gives us an expression for the *intensity*, I, of the wave—which of course is what is actually measured in an experiment. Thus

$$I = A'^2 \Sigma_{\mathbf{r}'} \exp\{i(\mathbf{k}_0 - \mathbf{k}).\mathbf{r}'\} \Sigma_{\mathbf{r}''} \exp\{-i(\mathbf{k}_0 - \mathbf{k}).\mathbf{r}''\} \tag{15.8}$$

$$= A'^2 \Sigma_{\mathbf{r}'} \Sigma_{\mathbf{r}''} \exp\{i(\mathbf{k}_0 - \mathbf{k}).(\mathbf{r}' - \mathbf{r}'')\} \tag{15.9}$$

where we have now had to introduce a second set of vectors, $\mathbf{r}''$, which run over the same atomic positions as the $\mathbf{r}'$. This expression can be split into two parts—one in which the $\mathbf{r}' = \mathbf{r}''$ and the other for which $\mathbf{r}' \neq \mathbf{r}''$. The first part contributes unity for each atomic site and so it is equal to N, the total number of atoms. We then obtain

$$I = A'^2 [N + \Sigma_{\mathbf{r}'} \Sigma_{\substack{\mathbf{r}'' \\ \mathbf{r}' \neq \mathbf{r}''}} \exp\{i(\mathbf{k}_0 - \mathbf{k}).(\mathbf{r}' - \mathbf{r}'')\}] \tag{15.10}$$

The terms in the double summation will only contribute to I if, just as in the crystalline case, they are in phase and this will only occur if the products of each of the components $(\mathbf{k}_0 - \mathbf{k})_x (\mathbf{r}' - \mathbf{r}'')_x$ (and similarly for the y and z components) are each an integral number of 2π. (These are similar conditions to those which were necessary in the crystalline case, section 2.7, p. 25). In general these conditions cannot be satisfied, but if we consider an atom at a particular $\mathbf{r}'$ then, as already discussed, its close neighbours will be regularly spaced and hence their contributions to the sum in (15.10) for the first few atoms around that $\mathbf{r}'$ will be in phase if their orientation satisfies the Bragg relation. Beyond a certain distance the periodicity of the structure is lost and so the contributions of more distant atoms will, on average, cancel out. Thus there will be a diffracted beam from the atoms that are close to certain of the $\mathbf{r}'$ and its direction will depend on the orientation of the atoms around that $\mathbf{r}'$.

If we now consider the effects of atoms in the neighbourhood of an atom at another value of **r′**, they will also contribute a diffracted beam if they have a suitable orientation, but since the orientation of the atoms about the new **r′** will probably be different from that which we first considered, the direction of their diffracted beam will not be parallel to the first beam. Thus as we run over all possible values of **r′** we obtain a set of diffracted beams in different directions which are all at the same angle to the incident beam so that they form a cone of radiation. When this falls on to a photographic emulsion or some other detector normal to the incident radiation, it will produce a circular diffraction pattern.

The situation is analogous to the pattern of rings that is obtained from a crystalline powder (section 2.9, pp. 27–8). In the powder certain particles are oriented so that they contribute to a cone of radiation at a particular orientation to the incident beam, whereas in an amorphous material we can consider each atom with its near neighbours as being a 'particle' of a pseudo-crystalline powder, some of which will have the correct orientation to contribute to the diffraction pattern. The only difference is that even with a very fine crystalline powder the periodicity extends over a volume of millions of atoms. This produces very sharp, well-defined rings, whereas the number of atoms which contribute from a particular region of an amorphous material will be perhaps a hundred or even fewer, and so the rings are very broad with poor contrast (Fig.15.6). We emphasize that just as in the

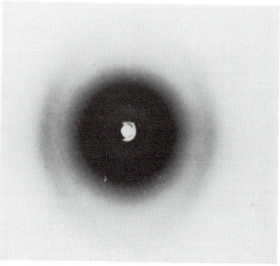

FIG 15.6. X-ray diffraction of an amorphous material (epoxy-resin) showing the broad, diffuse rings that are obtained. The circles are distorted because the film is mounted as an upright cylinder around the sample; hence the vertical distances are exaggerated. (Photo by D. Owen).

diffraction pattern from a crystalline powder, most of the particles do *not* contribute because they are not oriented at the Bragg angle to the incident beam, so in an amorphous material only certain regions will have the correct orientation for diffraction.

15.12 The radial (or pair) distribution function

It is now profitable to pursue the mathematical treatment of (15.10) a little further. We introduce the concept of a radial distribution function,† $P(r)$, which gives the probability that there will be an atom at a distance, r, from a particular atom. The general form of $P(r)$ is fairly clear. Since no two atoms can be closer together than an atomic diameter, $P(r)$ must be zero until r reaches the value of that diameter. If the material has some local order or periodicity of dimension a, then $P(r)$ will have maxima at $r = a$, $2a$ and possibly at other values as well. But at large values of r, because we have an amorphous material, there will be no preferred distances where atoms will be present and so $P(r)$ will flatten off to a constant value (Fig. 15.7) which for convenience we set at unity. (NB Most workers plot $4\pi r^2 P(r)$ in order to take account of the increase in the number of atoms as r increases.)

With this description of $P(r)$ we can now express the summation over $\mathbf{r}''$ in (15.10) as an integral of the form

$$n \iiint P(r) \exp \{(\mathbf{k_0} - \mathbf{k}) \cdot \mathbf{r}\} \, d\mathbf{r} \qquad (15.11)$$

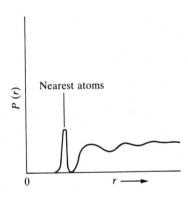

FIG 15.7. The radial (or pair) distribution function $P(r)$. The distances r from a given atom at which other atoms are most probably situated are indicated by the maxima in the curve. The very sharp peak at low r indicates the position of the nearest neighbours.

† Often abbreviated to RDF. $P(r)$ is sometimes called the *pair* distribution function.

where $r = (r' - r'')$ and n is the average number of atoms per unit volume. We see that the function $P(r)$ weights the contribution of the exponential at those values of r for which there is a greater chance of an atom being present, and indeed it plays a similar role to the periodic function $f(r')$ which we introduced into (2.4) in order to convert a summation into an integral. So, having written (15.11) as equivalent to the summation over r'', we now need to complete the evaluation of (15.10) by doing the second summation—that over all atomic sites r'. But clearly each site is equivalent and each will therefore give the same contribution. Hence the final result of the double summation will be $N \times$ (15.11). We can therefore write that the diffracted intensity is

$$I = A'^2 N[1 + n\iiint P(r) \exp\{i(k_0 - k).r\}\,dr].\qquad(15.12)$$
$$\text{volume}$$
$$\text{irradiated}$$

We can generalize this by noting that the constant value (unity) of $P(r)$ at large r does not contribute to the value of the integral because all values of the exponential will then be summed indiscriminately and hence they will cancel out—it is only periodic deviations of $P(r)$ from unity that will contribute. We can therefore rewrite (15.12) as

$$I = A'^2 N[1 + n\iiint\{P(r) - 1\} \exp\{i(k_0 - k).r\}\,dr].\qquad(15.13)$$
$$\text{all}$$
$$\text{space}$$

where we can now integrate over all space because $\{P(r) - 1\}$ becomes very small for large r and so it does not contribute to the integral as $r \to \infty$. It is useful to rearrange (15.13) as

$$\{I/(A'^2 N) - 1\}/n = \iiint\{P(r) - 1\} \exp\{i(k'.r)\}\,dr.\qquad(15.14)$$

where $k' = (k_0 - k)$. Now the integral in (15.14) is the Fourier transform of $P(r) - 1$ (section 2.6, p. 23), which in itself does not seem particularly interesting until we recall that the Fourier transform of the Fourier transform of a function is $8\pi^3$ times the original function—i.e. it *untransforms* the function (see Lipson and Lipson, chapter 3). Hence we reach the important conclusion that the Fourier transform of $\{I/A'^2 N) - 1\}/n$ is $8\pi^3 \times \{P(r) - 1\}$

i.e. $$\{P(r) - 1\} = (1/8\pi^3 n) \iiint\{I/(A'^2 N) - 1\} \exp i(k'.r)\,dk'.\qquad(15.15)$$
$$\text{all } k'$$

Thus by measuring the intensity distribution of the diffracted beam as a function of k', i.e. of the angle between incident and diffracted beams, we can derive the radial distribution function for the material.

It should be noted that the analysis which we have given only applies to a system of identical particles that have spherical symmetry, but the general form of the results is similar for more complicated systems and it is also valid for liquids and gases.

15.13 X-ray experiments

As an illustration of the technique we describe the pioneer work of Warren (1937) on silica-glass, SiO_2. Figure 15.8(a) shows a trace of the intensity of the diffracted beam as a function of scattering angle in which several maxima are clearly evident, corresponding to the dense part of the rings in a diffraction photograph. The Fourier transform of this trace was calculated and $P(r)$ is shown, multiplied by $4\pi r^2$, in Fig. 15.8(b). The peaks indicate the distances from a given atom at which it is most likely that another atom will be found. The first peak occurs at 0·162 nm — this is exactly the same distance as the Si–O separation in *crystalline* SiO_2, thereby demonstrating that the nearest neighbours of an atom in an amorphous material have the same separation as in a crystal. The area under the peak is a measure of the number of nearest neighbours. Warren showed that this corresponded to each Si atom being surrounded by four O atoms. Thus each Si atom is at the centre of a tetrahedron of O atoms, again just as in the crystalline form of SiO_2 (see Fig. 15.2(b)). The next peaks correspond to O–O and Si–Si pairs. The latter separation is found to be about twice the Si–O distance; this suggests that the Si–O–Si bonding is almost collinear. Beyond about 0·5 nm $P(r)$ becomes constant (note that the figure plots $4\pi r^2 P(r)$) because there is no long-range order. This is because the orientation of the Si–O tetrahedra changes through the material so that they are twisted with respect to one another.

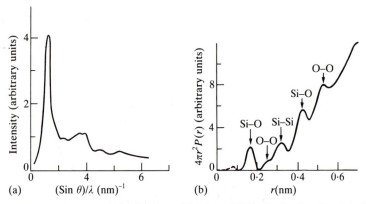

FIG 15.8. (a) The intensity distribution and (b) the radial distribution function derived from it for vitreous SiO_2. The peaks of $P(r)$ are labelled with the atomic pairs to which they relate (after Warren, *J.Appl.Phys.* **8**, 646 (1937).)

$P(r)$ for models such as those described in section 15.5 can be calculated and then compared with experimental vales of $P(r)$ for real samples (see Fig. 15.9). In this way the validity of the representation of a particular material by a certain model can be evaluated.

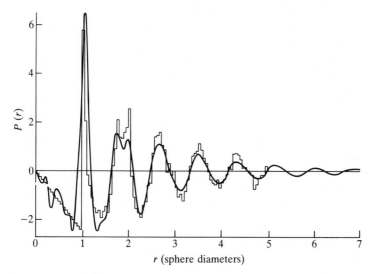

FIG 15.9. The reduced radial distribution function derived from measurements on amorphous $Ni_{76}P_{24}$ (smooth curve) compared with that calculated for a model of dense random packing of spheres. (Cargill and Finney, *Solid St.Phys.* **30**, 304 (1975).)

15.14 Electrons in amorphous materials

In Chapter 8 we described the properties of electrons in a perfectly periodic lattice—the electron wave functions are continuous throughout the material and their permitted energies span certain energy bands that are separated from one another by gaps of forbidden energy. Electrons within the bands have a high mobility. They can travel freely through the material and they only undergo scattering (which gives rise to electrical resistance) if there are faults in the lattice periodicity. In semiconductors the presence of impurities introduces electron states—i.e. donor and acceptor levels—into the band gap, and electrons in these states are *localized*. They can only contribute to the conductivity by being thermally excited into a higher band. This is the principle of doping (section 9.6, p. 147).

How is the band model affected if, as in amorphous materials, we have a potential of random periodicity? At first sight one might think that the whole band picture must be discarded, but there are two common observations which suggest that the main features must still be preserved. These are (i) ordinary window glass is transparent to visible light and hence it must have a band gap of several eV, otherwise the photons would be absorbed and it would appear opaque, (ii) liquid mercury (and other molten metals), even though it must have a random atomic arrangement, is a good electrical conductor—hence the electron mobility is still quite high.

The reason why the band picture is still valid is that the major influence on an electron is *short range order* and this, as we have seen, is still present in amorphous materials. The importance of short range order is surely reasonable since an electron is much more likely to be affected by the disposition of the atoms that are close to it than by those that are more distant. This local periodicity still gives rise to permitted and forbidden bands of energy. However the absence of long range order has the effect of *smearing out* the band edges into tails (Fig. 15.10). This is analogous to the diffraction patterns that we described in section 15.11. These were smeared-out versions of those that would be obtained from crystalline specimens. A consideration of the consequences of the *tight binding model* (section 8.19, p. 140) in which the electron levels of neighbouring atoms shift with respect to one another as the atoms are brought closer together, would also suggest that bands should occur in amorphous systems.

The electron states in the band tails are not the same as those within the bands. The tail states are *localized* and hence electrons in those states have a very low mobility compared with those in the bands. Their presence is very important for explaining the metal–insulator transition (section 15.17), and we shall leave further discussion to later sections.

15.15 The electrical conductivity of amorphous metals

Just as in crystalline materials, it is the position of the Fermi energy, E_F relative to the energy bands that determines whether a material will exhibit metallic or semiconducting (or insulating) behaviour. If E_F is within an energy band we have a metal, whereas for a semiconductor E_F is in the band gap (Fig.15.10). Since, as we have just described, the general form of the bands is maintained even in the amorphous state, it is not surprising that the position of E_F will also be similar to that in the crystalline material. Hence in amorphous metals E_F will lie within the conduction band (see Problem 15.4) and so there will always be electrons occupying extended states which can take part in the transport of charge. Nevertheless the disorder is usually so great that the electron mean free path is very small indeed and it is often only of the order of the interatomic spacing. This gives rise to values of the resistivity which are very much higher than those measured in crystals. For such strong scattering the division of the resistivity into two individual components—one due to defects and the other due to phonon scattering (as in section 8.3, p. 126) is not satisfactory. The disorder due to the amorphous

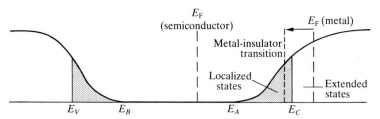

E_F
(semiconductor)

E_F (metal)

Metal-insulator transition

Localized states

Extended states

E_V E_B E_A E_C

FIG 15.10. Electron states for amorphous materials. The band edges are smeared into *tails*, where the states (shaded) are *localized*. In a metal E_F lies within the extended states. In a semiconductor it will be around the middle of the band gap. In a metal–insulator transition E_F shifts from the region of extended states to that of the localized states. The region between E_v and E_c is called the *mobility gap*.

nature of the material and that occasioned by the thermal motion cannot be separated. Nevertheless the resistivity does change slowly with increasing temperature. It usually increases, although in some materials it has been observed to decrease slightly.

15.16 The minimum metallic conductivity—electron localization

Experiments on a range of amorphous alloys suggest that the electrical conductivity at low temperatures, for materials showing metallic behaviour, does not seem to go below a temperature-independent value of about $5 \times 10^5 \, \Omega^{-1} \, m^{-1}$. Lower conductivity values are only observed in materials that exhibit semiconducting behaviour. This minimum metallic conductivity can be used to calculate the electron mean free path, l (e.g. by using eqn (8.2), p. 125) and it is found this is then about an atomic spacing, a. Once the mean free path has been reduced to this value our ordinary concept of electron diffusion and scattering must be revised and it has been proposed that the condition† $l = a$ marks the change from extended electron states (which give rise to ordinary metallic conductivity) to localized electron states, which we shall consider in the following sections. Using eqn (8.2) (p. 125) it is quite straightforward to show (see Problem 15.5) that the minimum metallic conductivity is of the order of $e^2/(3\hbar a)$.

15.17 The metal–insulator transition

If the Fermi energy lies just above the boundary between the localized and the extended states it is possible, by increasing the disorder, for the range of localized states to be extended beyond E_F. Conversely, but with a similar effect, E_F can be shifted into the localized region. This can be done in several

---

† This is a simplification of the Ioffe–Regel criterion, which states that localization occurs when $kl \sim 1$, where k is the wave vector.

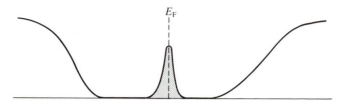

FIG 15.11. In amorphous semiconductors it is suggested that there is a high concentration of localized states near the centre of the gap that anchors E_F in that region.

ways, e.g. by changing the alloy composition, by applying an electric or a magnetic field, or by a mechanical stress. Electrons in the localized states can still contribute to the conductivity but this is controlled by hopping mechanisms, which will be described in section 15.19. This conductivity is very small and so the material becomes an insulator. These transitions, in which E_F crosses the boundary between the extended and the localized states, are sometimes called *Anderson transitions*. They can also be observed in certain crystalline alloys in which there is a random disposition of one of the components; this can therefore be considered to be an amorphous phase.

15.18 Amorphous semiconductors

In amorphous semiconductors E_F must lie somewhere within the band gap (otherwise the material would be metallic) and experimental evidence suggests that it is pinned in the region of the midpoint of the gap. The reason for this is not completely established, but it seems likely that a high density of deep levels due to impurities or defects will tend to anchor E_F (Fig. 15.11) in a similar way to that which causes E_F to lie close to the impurity levels in doped crystalline materials (section 9.10, p. 153).

We have already mentioned that the electron states within the tails of the valence and conduction bands are localized. It will be noted that these states bear a resemblance to the set of impurity levels that are produced close to the band edges in heavily doped semiconducting crystals (section 10.11, p. 172). But in these crystals it is the impurity atoms that can be considered to upset the lattice periodicity and that give rise to the localized states, whereas it should be emphasized that in amorphous materials no impurities are necessary. In fact until recently it was found to be rather difficult to dope (section 9.6, p. 147) amorphous semiconductors by the addition of other elements. This is because disorder and defects (section 15.6) are able to accommodate any additional bonds required for the extra electrons (or holes) on the impurities and so there are no loosely bound electrons (holes) remaining that would give rise to donor or acceptor levels (see Fig. 9.4, p. 149). Doped amorphous silicon is dealt with in section 15.20.

Let the energies at which the change from localized to extended states occurs be E_c and E_v in the conduction and valence bands respectively, and let E_A and E_B be the ends of the tails in each band (Fig. 15.10). Since the electron mobility in the localized states is very small, E_c and E_v are sometimes called the *mobility edges* or *shoulders*, and the region between E_c and E_v is referred to as the *mobility gap*.

15.19 Conductivity mechanisms

How will such an arrangement of localized and extended states influence the temperature dependence of the electrical conductivity? We shall discuss this in terms of electron conductivity but, as in chapter 9, analogous effects will also occur for holes. We need to consider four mechanisms:

(a) At high temperatures for which the thermal energy is sufficient to excite electrons to the extended states beyond E_c in the conduction band (this will be at room temperature for Ge and Si), 'ordinary' conductivity due to the electrons drifting in the electric field will occur. This will be of the form (ignoring the $+1$ in the Fermi function)

$$\sigma = \sigma_0 \exp\{-(E_c - E_F)/(kT)\} \tag{15.16}$$

where σ_0, a function of the mobility and the density of states, will be the minimum conductivity discussed in section 15.16.

(b) At lower temperatures excitation can only occur to the localized states between E_A and E_c. Electrons in these states can contribute to the conductivity only if they are able to acquire sufficient energy, ΔW_1, to hop from one localized site to another. Thus the minimum activation energy will be $E_A + \Delta W_1$ and the conductivity will be of the form

$$\sigma = \sigma_1 \exp\{-(E_A + \Delta W_1 - E_F)/(kT)\} \tag{15.17}$$

where σ_1 will be much less than σ_0 in (15.16)—about one thousand times smaller—and it will depend on the fluctuations in the electron motion due to phonon interaction, i.e. on the rate at which an electron tries to jump. This process is called *thermally assisted hopping*.

(c) At still lower temperatures two other hopping processes are still possible, provided that, as discussed in section 15.18, a band of localized states exists around E_F (Fig. 15.11). Just as in (b), thermally assisted hopping to neighbouring sites can still occur from states that are close to E_F. This will give a conductivity of the form

$$\sigma = \sigma_2 \exp\{-\Delta W_2/(kT)\} \tag{15.18}$$

where ΔW_2 is the average energy difference between sites. This will be approximately one half of the energy spread of the band around E_F. σ_2 will depend, as did σ_1, on the electron jump frequency.

(d) At very low temperatures, where $kT \ll \Delta W_2$, the rate of thermally assisted hopping to neighbouring states will be very small and it is more likely that an electron will be able to find a state of similar energy by tunnelling beyond its nearest neighbours to more distant sites, since then there will be a greater selection of possible energies. This is called *variable range hopping* and it leads to a conductivity with a temperature dependence (unlike that of processes (a) to (c)) of the form $\exp(-B/T^{\frac{1}{4}})$. This mechanism, which was first proposed by Mott, can be derived by the following simple argument.

Since the electron states around E_F are localized, their wave functions will fall off exponentially with distance R as $\exp(-\alpha R)$, where α is a constant. Hence the probability of hopping to R will vary as

$$\exp(-2\alpha R). \tag{15.19}$$

To derive the hopping energy we note that the minimum energy between states around E_F can be derived from the density of states per unit volume, $G(E_F)$ as follows. The number of states between E and $E + dE$ in a volume V is

$$V\, G(E_F)\, dE. \tag{15.20}$$

If V contains just one state we can set (15.20) equal to unity and dE will then be the energy spread between one state and the next. Hence the average separation between neighbouring states, ΔW_3, is

$$\Delta W_3 = 1/\{V\, G(E_F)\}. \tag{15.21}$$

If the electron is able to sample states within a radius R this becomes, on putting $V = 4\pi R^3/3$,

$$\Delta W_3 = 3/\{4\pi R^3\, G(E_F)\}. \tag{15.22}$$

Thus the farther an electron is able to tunnel, the greater will be the likelihood that it will find a site with small ΔW_3. The probability for such a thermal excitation, ΔW_3, will be $\exp\{-\Delta W_3/(kT)\}$ and this, together with (15.19) gives a combined probability of the form $\exp[-2\alpha R - \{\Delta W_3/(kT)\}]$. On substituting for ΔW_3, from (15.22) this yields

$$\exp[-2\alpha R - 3/\{4\pi R^3\, G(E_F)\, kT\}]. \tag{15.23}$$

The maximum probability will occur when the magnitude of the exponent is a minimum i.e. when

$$2\alpha - 9/\{4\pi R^4\, G(E_F)\, kT\} = 0 \tag{15.24}$$

hence

$$R = [9/\{8\pi\alpha G(E_F)\, kT\}]^{\frac{1}{4}} \tag{15.25}$$

or

$$R \approx 1/\{\pi\alpha G(E_F)\, kT\}^{\frac{1}{4}}. \tag{15.26}$$

On substituting for R in (15.22) we find that the conductivity is of the form

$$\sigma = \sigma_3 \exp(-B/T^{\frac{1}{4}}) \qquad (15.27)$$

where

$$B = 2 \cdot 75[\alpha^3/\{\pi G(E_F)k\}]^{\frac{1}{4}}. \qquad (15.28)$$

It is difficult to get good estimates for α and $G(E_F)$ but experiments yield a value for B of about 2. σ_3 will be controlled by the electron jump frequency. To check the $T^{\frac{1}{4}}$ term with any precision is, of course, rather difficult, but there is now substantial evidence to confirm the form of (15.27); see Fig. 15.12.

Taking account of all these mechanisms we see that a plot of $\log \sigma$ against $1/T$ should show four distinct regions if all the processes (a) to (d) are operative. Three of these, corresponding to (a), (b), and (c) should be straight lines† with slopes that are proportional to $(E_c - E_F)$, $(E_A + \Delta W_1 - E_F)$ and ΔW_2 and this is illustrated by the set of results in Fig. 15.13. If variable range hopping occurs, it will produce a slight curvature at large values of $1/T$ because of the $T^{1/4}$ dependence.

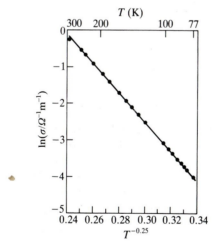

FIG 15.12. Variable range hopping. Experimental data for an amorphous Ge film which demonstrates the $T^{1/4}$ dependence of the logarithm of the electrical conductivity, σ. (Gilbert and Adkins, *Phil.Mag.* **34**, 147 (1976).)

† This assumes that E_F and the band parameters E_c, E_v, E_A, E_B, do not change with temperature.

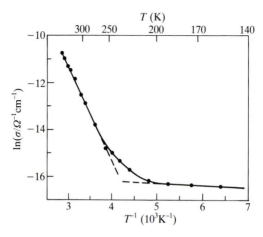

FIG 15.13. Experimental data for amorphous $(As_2Se_3)_{0.95}Tl_{0.05}$ showing three distinct regions of conductivity behaviour, which are thought to correspond to the three conductivity mechanisms (a) to (c) discussed in the text. (After Kotkata, El-Fouly, Fayek and El-Hakim, *Semicond. Sci. Technol.* **1**, 315 (1986).)

15.20 Doped amorphous silicon

Amorphous silicon is usually made by vapour deposition of Si on to a hot ($\sim 240°C$) substrate from the breakdown of silane (SiH_4) or disilane (Si_2H_6) by an r.f. glow discharge (or sometimes by ultra-violet). By adding either diborane (B_2H_6) or phosphine (PH_3) to the silane p or n type material is produced. The precise mechanism for this is still not established with certainty, but it is thought that it is made possible because the hydrogen from the silane renders inactive any defect sites, (e.g. dangling bonds, section 15.6) that would otherwise trap the doping atoms. These atoms are now able to enter substitutionally for Si and they should be fourfold coordinated, but just as in crystalline materials (section 9.6, p.147), the deficit or excess of electrons beyond four in the outer shell leads to the formation of shallow electron or hole states.

The ability to fabricate doped amorphous material enables it to be used in devices which rely on the properties of a p–n junction. At present the most important application of doped amorphous Si is in solar cells (section 10.18, p. 183). These need to have large surface areas in order to trap as much light as possible but with Si single crystals the size is very severely restricted, whereas the amorphous material can be made in quite large sheets. For this reason another potential application is in switching arrays for large area liquid crystal displays.

15.21 The Hall effect

In section 9.8, p. 150, we showed that a measurement of the Hall co-efficient, R_H, enabled us to ascertain the sign of the dominant carriers in a semiconductor. Positive and negative values of R_H corresponded to conduction by holes and electrons respectively. In amorphous semiconductors this is not so. R_H usually has a sign which is opposite to that which we would expect—so n-type material would have a *positive* R_H value.

The simple arguments of section 9.8 break down because they assume carrier diffusion and weak scattering whereas, as we have seen, we are now concerned with hopping mechanisms and a very short mean free path. The simple treatment is no longer appropriate and as yet there is no generally agreed theory of the phenomenon.

15.22 Photoconductivity

When photons with an energy equal to or greater than the band gap fall onto an amorphous semiconductor they are absorbed, and an electron–hole pair is formed. This is the same effect as we have discussed in section 9.16, p. 156, except that in the amorphous case there is no requirement for wave vector conservation. The extra number of charge carriers will increase the conductivity of the material, but the details of the process will depend very much on the probability for carrier trapping and electron–hole recombination, since these processes will limit the actual numbers of carriers that are available for charge transport.

15.23 Xerography

One of the most widespread applications of the photoconductive effect in amorphous semiconductors (Se) is the xerographic process for photocopying documents. The essential steps in the process are shown in Fig. 15.14. An

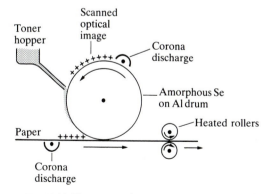

FIG 15.14. The xerography process—see text.

amorphous Se film, which has been deposited on an earthed Al drum, is charged in a corona to about 700 V. As the drum rotates it is exposed to a scanned optical image of the document and wherever the Se is illuminated photons are absorbed, each creating an electron–hole pair. The electrons *cancel out* the positive surface charge and the holes are attracted to the earthed base. Hence the Se surface now possesses an electrical image of the document, in which the dark areas are positively charged and the bright areas are uncharged. The Se is then dusted with negatively-charged toner (carbon) particles, which are each encapsulated in a low melting-point plastic shell. These will be attracted to the positive (dark) regions. The toner is transferred to a positively-charged sheet of paper which passes through heated rollers to melt the plastic coating of the carbon particles, thereby fixing the image on the paper.

In order for the process to work satisfactorily the Se must have a very high electrical resistivity so that the charges which remain in the unexposed areas do not leak away. It must not contain any traps due to defects or impurities which would prevent the photo-holes from reaching the earthed substrate. In these respects amorphous Se is far superior to the crystalline material. It is also much more efficient in electron–hole pair production at the frequency of ordinary light sources.

A similar application of the photoconductivity of Se is in the vidicon tube of a television camera.

15.24 Superconductivity

Some amorphous materials exhibit superconductivity. This is discussed in section 14.26, p. 265.

15.25 The thermal properties of amorphous materials

What is the spectrum of thermal vibrations in amorphous materials? We need to know this in order to discuss their thermal properties—and especially their specific heat and thermal conductivity. In marked contrast to some of the other properties discussed in this chapter, the thermal properties of amorphous materials are almost independent of the actual substance under investigation.

15.26 The specific heat

In sections 5.7 to 5.12 (pp. 83–8) we derived expressions for the thermal energy and the specific heat of crystals using the Debye theory. The essence of that theory is that the material is considered to be an elastic continuum, i.e. no assumption is made about the positions of the individual atoms. The predictions of the theory should therefore apply equally to amorphous and to crystalline solids and by and large this is indeed found to be so. At high temperatures the specific heat is constant—equal to the Dulong–Petit value of $3R$ (section 5.12, p. 88) and as the temperature is reduced it follows the Debye curve until at low temperatures the specific heat varies as T^3. However, whilst this T^3 relationship is indeed observed down to about 2 K, below this temperature the dependence changes and it is found that the specific heat, C, is dominated by a term that is linear in T. Thus in the liquid helium region C is of the form

$$C = AT + BT^3, \qquad (15.29)$$

so that a plot of C/T against T^2 should be a straight line (Fig.15.15). It is remarkable that the relationship (15.29) is found in all non-metallic amorphous materials that have been investigated, be they glasses, plastics, or semiconductors (the only exception appears to be amorphous arsenic, which does not seem to have the AT term). The explanation of this linear term in such a wide variety of materials has been an intriguing challenge to theoreticians. Do amorphous metals show similar behaviour? The problem is that a linear specific heat is the hallmark of a system of quasi- free electrons (section 7.8, p. 113) so a metal always has a specific heat of the form of (15.29). Nevertheless experiments on amorphous superconductors at very low temperatures (where the electronic contribution is negligible) also indicate that the specific heat has an additional contribution that is linear in T.

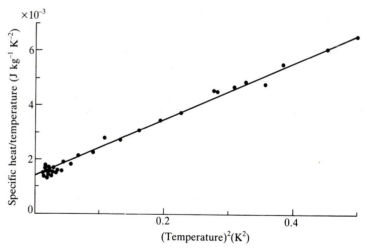

FIG 15.15. The specific heat of amorphous B_2O_3 plotted as C/T against T^2, to demonstrate the validity of eqn (15.29). (Stephens, *Phys.Rev.* **B13,** 858 (1976).)

15.27 The two-level-state model

The model that is now widely accepted is that the additional specific heat is due to the excitation of what have been termed localized two-level states (TLS) in the material. The precise origin of these states has not been firmly established, but a simple geometrical model can be suggested. In an amorphous solid, unlike in a crystal, the atoms have no 'correct' positions and it is quite possible that there are two atomic configurations that differ by only a very slight displacement or rotation and by a correspondingly small energy, E (Fig. 15.16). However, because the material is amorphous, these two possible arrangements will not be precisely the same at all atomic sites and so the values of E will be slightly different from one another. At 0 K the lower of these two states will be occupied, but as the temperature is raised, excitation will occur to the higher state. Because there is a spread of values of E this will not give a peak in the specific heat (which is what would be observed if all the E values were the same) but there will be a continuing addition to the Debye specific term as the temperature is raised and more TLS are excited.

The calculation for this additional specific heat is quite straightforward. We assume that each site of a TLS is isolated from the others and that the energy of the upper state is E above the lower one, which we take to be zero. At a temperature T the mean energy, $\langle E \rangle$, is then

$$\langle E \rangle = E \exp\{-E/(kT)\}/[1 + \exp\{-E/(kT)\}]. \tag{15.30}$$

The specific heat is obtained by differentiating (15.30) with respect to T, i.e.

$$(\partial \langle E \rangle / \partial T)_v = (E^2/kT^2) \exp\{-E/(kT)\}/[1 + \exp\{-E/(kT)\}]^2. \tag{15.31}$$

This is the specific heat for a single site and we can obtain the total specific heat by summing over all sites. We make the simplifying assumption that the energy of the sites is distributed evenly over the entire energy range, i.e. the

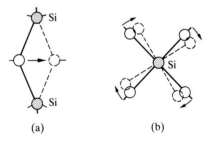

(a)   (b)

Fig 15.16. Two possible atomic configurations for two-level states in SiO_2. (a) the O atom can jump from left to right; (b) the arrangement of four O atoms can rotate slightly around the Si atom.

density of states, G_{TLS} is a constant. The total contribution to the specific heat due to the TLS (C_{TLS}) will be obtained by multiplying (15.31) by G_{TLS} and then integrating over all E; hence

$$C_{TLS} = G_{TLS} (1/(kT^2)) \int E^2 \exp\{-E/(kT)\}/[1 + \exp\{-E/(kT)\}]^2 dE. \qquad (15.32)$$

If we set $x = E/(kT)$ this becomes

$$C_{TLS} = G_{TLS} k^2 T \int x^2 e^{-x}/(1 - e^{-x})^2 \, dx. \qquad (15.33)$$

The integral has a value $\pi^2/6$ and so

$$C_{TLS} = (\pi^2/6) G_{TLS} k^2 T. \qquad (15.34)$$

Thus the additional specific heat is proportional to T, as is observed. It should be noted that this linearity is dependent on the assumption that the density of states, G_{TLS} is constant.

15.28 The thermal conductivity of amorphous materials

Just as the specific heat of amorphous materials showed unexpected behaviour, so does the thermal conductivity, κ. In general κ is very small at room temperature and it is further reduced on cooling. This low conductivity is a consequence of the atomic disorder, which is responsible for the very small mean free path, l, both of the electrons (in metals) and of the phonons.

In section 6.2, p. 96, we introduced the kinetic theory expression for κ

$$\kappa = c_v \, lv \qquad (15.35)$$

where c_v is the specific heat per unit volume of the heat carriers and v is their velocity. Since this is a completely general formula, it should apply to amorphous materials. In view of the atomic disorder the mean free path, both for the electrons (in metals) and for the phonons, will be small and this is confirmed by the low values of κ that are actually measured. In insulators if the value of l is calculated using measured values of the other quantities, it turns out to be surprisingly small—very often less than an atomic spacing. We are then confronted with the problem of what is meant by the mean free path of a wave which is smaller than the spacing of the discrete elements which make up the wave. This can get us into very deep water and the situation is very similar to that which we discussed with reference to electrons in section 15.16.

There are two important features that are observed in practically all amorphous substances:

(a) As the liquid helium region is approached the conductivity stops decreasing and depending on the material, between, say, about 15 and 3 K (depending on the material) it becomes constant (Fig. 15.17), i.e. it passes through a plateau region.

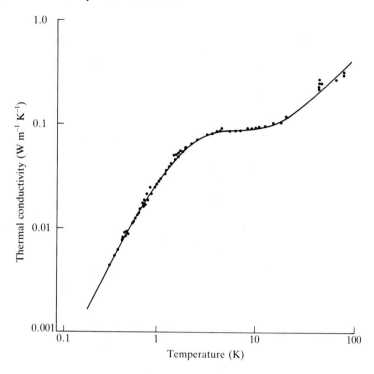

F𝐈G. 15.17. The thermal conductivity of an epoxy-resin as a function of temperature (log–log plot), which demonstrates the T^2 behaviour at the lowest temperatures followed by the plateau in the liquid helium region (C. I. Nicholls, D.Phil. thesis, Oxford, 1982).

(b) At still lower temperatures it decreases again with a temperature dependence which is approximately T^2.

The T^2 dependence at the lowest temperatures is caused by phonon interactions with the TLS, discussed in section 15.27. Since each of these states is localized to a particular part of the material, they cannot themselves transport energy, but each can absorb and re-emit a phonon of energy E, if that E corresponds to the energy difference between its two possible states.† Thus the presence of the TLS provides a mechanism for phonon scattering.

Since the specific heat of phonons varies as T^3, a T^2 dependence for κ would suggest from (15.35) that the mean free path should vary as T^{-1} (since v is constant). A full derivation for l is beyond the scope of this book but its

---

† This is called resonant absorption. Non-resonant interactions are also possible, but they are not so important.

proportionality to T^{-1} is easily demonstrated. A phonon of energy E is absorbed each time a two-level centre is excited into its upper state. The transition rate for this process will be proportional to the number of TLS with energy E (which we have already assumed to be a constant) and also to the square of the phonon field,† i.e. to their energy E. But phonons of this energy are dominant at a temperature $T = E/k$ (section 6.4, p. 97) and since the phonon mean free path must be inversely proportional to the phonon absorption rate, i.e. to E^{-1} (or T^{-1}) the mean free path should vary as T^{-1} as the experimental observations require.

15.29 The thermal conductivity plateau

The plateau region that occurs at slightly higher temperatures is believed to arise because as the temperature is raised, thermal vibrations of higher energy (and hence of longer wave vector, $\mathbf{k}$) are excited, and, as in section 15.16, if l for these excitations decreases to a value of the order of the interatomic spacing, then, instead of behaving like phonons, i.e. *travelling waves*, which are able to transport the heat along the sample, they are instead, localized. If this localization does indeed occur then these excitations cannot transport thermal energy and this will give rise to a blockage in the heat transport mechanism—i.e. to the plateau region, which we have described. There is still considerable debate as to why in amorphous materials (but not in crystals) the high energy thermal excitations should be localized, and the situation has as yet not been completely resolved.

15.30 Ultrasonic attenuation by two level states

The most convincing evidence for the existence of two level states (TLS) (section 15.27) are the measurements of ultrasonic attenuation in glasses at temperatures below 1 K. At *low* acoustic intensity the ultrasonic attenuation increases as the temperature is reduced and it is suggested that this is due to resonant scattering of the ultrasonic phonons by the TLS, more of which will be in their ground state at low T. However, as the intensity is increased, the attenuation is considerably reduced (Fig. 15.18). This reduction is interpreted as being caused by the ultrasonic phonons saturating the TLS so that they are all excited to their upper state. When this occurs (provided that the lifetime in the upper state is sufficiently long) no more phonons can interact with the TLS and so there is less attenuation

15.31 Magnetic properties

We have already described in sections 12.3 and 12.4 (p. 201–2) that magnetic ordering is due to the *exchange interaction* between neighbouring ions and this in turn is dependent on the extent and overlap of the ionic wave

---

† From time-dependent perturbation theory.

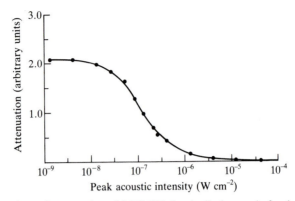

FIG 15.18. Ultrasonic attenuation of 0.592 GHz longitudinal waves in fused silica at 0.023 K. The attenuation *decreases* as the ultrasonic power is increased because the two-level states are saturated and this reduces the possibility of scattering. (After Golding, Graebner and Schutz, *Phys.Rev.* **B14**, 1660 (1976).)

functions. Such an overlap can of course still arise in the amorphous state, although it will not now be exactly the same for all ion pairs. Hence ferromagnetism can occur at and below a fairly well-defined Curie temperature and the general form of the phenomenon is the same as in crystals. True anti-ferromagnetism (section 12.13, p. 209) however, is *not* observed because this would require two specially preferred directions and the precise cancellation of two equivalent sub-lattices, but partial cancellation to give ferrimagnetism is possible.

The most important ferromagnetics which have been made are alloys† of the transition metals Fe, Co, Ni, together with a 'metalloid', usually C, B, P, or Si, with the transition metal component usually comprising about 80 per cent of the mixture. The simplest of these is $Fe_{80}B_{20}$, but many materials, some with as many as five ingredients, have been produced.

In some respects the characteristics of amorphous ferromagnets are not as good as those of their crystalline counterparts. Their Curie temperatures tend to be lower—from 200 to 400°C—compared with 700°C for transformer steel, and they are not so sharply defined. The maximum saturation magnetization is about 1·6 T compared with about 2 T for transformer steel. They are, however, mechanically very tough, they do not deform plastically and they have a very high electrical resistivity. They can be produced cheaply as continuous strip, wire, or ribbon which requires no further treatment.

† We do not discuss the properties of *spin glasses*. This term is usually applied to dilute alloys (crystalline and amorphous) that contain small quantities of magnetic ions; e.g. CuMn1 per cent, AuFe1 per cent (both crystalline), $(Fe_{0.64}Mn_{0.36})P_{16}B_6Al_3$, AlGd37 per cent (both amorphous).

15.32 Saturation magnetization; spin waves

At 0 K magnetic order is complete and the value of the saturation magnetization can be used to calculate the effective magnetic moment per magnetic atom. In pure crystalline Fe this is about 2·2 Bohr magnetons per atom.† In the amorphous alloys this is reduced and it is suggested that this occurs because electrons from the metalloid elements contribute to the 3d band and this will tend to reduce the number of unpaired electrons. Hence the effective magnetic moment will be lower.

As the temperature is increased the thermal energy misorients the elementary dipoles so that, just as in crystals, the saturation magnetization is reduced. It has a similar form, with a $(1 - aT^{3/2})$ dependence (eqn 12.5)), which is the relation which can be deduced from spin wave theory.

15.33 Magnetic anisotrophy

Because an amorphous material is homogeneous it should have no preferred directions and hence these ferromagnets should not exhibit any anisotropy, i.e. there should be no special 'easy' or 'hard' axes for magnetization (section 12.8, p. 205). Whilst this is more or less true, internal strains are always introduced during the rapid quenching involved in manufacture (all the materials are produced by the 'splat cooling' process (section 15.3), and even careful annealing introduces some alignment since this must always be done in a weak (e.g. the earth's) magnetic field. Nevertheless, although perfect anisotropy cannot be achieved, the energy required to change the direction of magnetization is very small. In addition, the homogeneity of the material and the absence of grain boundaries and of other structural defects reduce the possibility of pinning the domain walls. Hence these materials are in general magnetically soft, they show little magnetic hysteresis, and they have very narrow $B \sim H$ curves (Fig. 12.7, p. 208).

This makes amorphous ferromagnets an attractive proposition for use in power transformers and machinery, since in addition to the low hysteresis, their inherently high electrical resistivity (section 15.15) reduces the eddy current losses. This makes them compare favourably with grain oriented silicon-iron, the material which is in common use.

15.34 Magnetostriction

This is the change in dimensions when a magnetic field is applied. It is observed in most amorphous ferromagnets and its magnitude is very often similar to that of crystalline materials. However the effect is very small in certain glassy alloys that are rich in Co and this gives rise to a very low permeability because (section 12.9, p. 206) there will only be a very small change in elastic energy when a magnetic field is applied or is varied, since the

---

† See section 12.3, p. 201.

resistance to domain wall motion will be reduced. Such materials are very useful for magnetic screening. Conventional screening alloys, such as Supermalloy, are mechanically very soft and their magnetic properties degrade after deformation unless they are given further heat treatment. Amorphous alloys are very hard, they deform elastically, and their properties are unaffected by deformation.

15.35 Percolation theory

The principle of *percolation* is used in many theories that relate to the properties of irregular and amorphous systems. As a simple example let us consider a composite material consisting of a matrix (an electrical insulator) in which is embedded a certain volume fraction of identical metal spheres. What is the minimum volume fraction of spheres (which we shall call the fractional site occupancy) necessary to ensure that there will be an electrically conducting path between opposite sides of the material? The answer is approximately 0·2. The derivation of this value is the heart of the percolation problem.

The primary task is to calculate, for a certain fractional site occupancy, the probability that there will be a continuous path through the system, from one occupied site to its neighbour. If there is no neighbouring occupied site, the path is blocked and percolation cannot proceed. The value of the concentration of occupied sites to give a probability of unity, i.e. the certainty of a percolating path, is called the critical concentration, p_c.

These considerations are of course very relevant when we are considering any conductivity mechanism that involves an electron hopping from one site to another, or the propagation of atomic excitations through an irregular network. As p_c is approached there will be a rapid increase in conductivity which will flatten off as more conducting paths (in parallel) are formed.

There are two main types of percolating system:

(a) *Site percolation*, in which some sites are occupied and others are vacant (this is the type which we have already described). The path then has to go from one occupied state to another which is a neighbour. This is illustrated for a square network in Fig.15.19 in which the near neighbours have a common *side*.

(b) *Bond percolation*, in which all the sites are occupied but they are connected to one another by a certain number of bonds (or channels), some of which will be broken (or blocked). Percolation then occurs from one site to another via unbroken bonds.

The actual value of the critical concentration, p_c, depends on the dimensionality of the system and on the maximum number of neighbouring sites for (a), or of permitted bonds for (b); however it should be evident that as the value of any of these quantities is increased the chances of making a percolating path are improved and this will lower the value of p_c. In general p_c cannot be derived analytically and it is only with the advent of fast

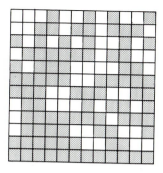

FIG 15.19. Percolation. A random square array generated by a computer (65 per cent of the sites are occupied). This shows several percolating paths between opposite sides. Paths must only cross between sides which are in contact, not between touching corners.

computers that reliable values of p_c have become available (see Problem 15.13). However, the important feature that is demonstrated by any system is that as the concentration of sites (or bonds) is increased, the length of the percolating path first increases quite slowly, but as p_c is approached the length increases very rapidly indeed (Fig. 15.20). This type of behaviour is very similar to the increase in correlation observed in critical phenomena, such as the onset of ferromagnetism or the change from a vapour to a liquid. For this reason the techniques of percolation theory are being introduced into these and similar fields.

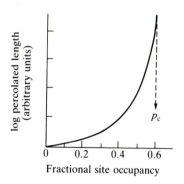

FIG 15.20. Percolation. The increase in the percolation length as a function of the site occupancy for the system of Fig. 15.19. This shows the rapid increase in length close to the critical concentration, p_c. This behaviour is similar to that observed in various types of critical phenomena.

If the percolation problem is for a classical model, direct contact between neighbouring sites (as already discussed) is required, but in many cases of interest there is the possibility of quantum-mechanical tunnelling not only between nearest sites but also between those that are not nearest neighbours. In such cases we must take account of the overlap of wave functions between the sites. If tunnelling is permitted between more distant sites then p_c will be smaller. Similar considerations will apply in substitutionally diluted magnetic systems to determine whether they will show magnetic ordering. We would then need to calculate the exchange interaction (section 12.3, p. 201) within the percolating array of magnetic atoms.

Problems

15.1 The mean distance which a heat pulse travels into a material of thermal diffusivity, D, in a time τ is of the order of $(D\tau)^{1/2}$ (compare with section 10.7, p. 169). Show that if a material is quenched from the melt at a temperature T_m and at the critical cooling rate r, then the maximum thickness that an amorphous sample can have will be $(DT_m/r)^{1/2}$.

Calculate the value of this thickness for a material that melts at 1200 K, has a thermal conductivity of 3 W m^{-1} K^{-1}, a specific heat of 6500 J K^{-1} m^{-3}, and is cooled at 10^{-6} K s^{-1}.

15.2 Produce a random network of close circles as follows. Draw a network of $9 \times 9 = 81$ squares, each of side 10 mm, and mark the centre of each square as an origin. These represent the centres of a perfect array of circles. Use a random number generator to produce a set of random numbers in the range -5 to $+5$. As the numbers are generated use the first as the x-coordinate and the second as the y-coordinate (in mm) and mark the centre of each misplaced circle, for each square in turn. Now start with a point near the centre of the array as $r = 0$ and draw a series of circles whose radii increase in steps of, say, 2·5 mm. By counting the number of points which fall between the sets of circles, plot out the radial distribution function.

This model is not entirely satisfactory because each origin is random with respect to the perfect array rather than to the neighbouring points. Devise a computer programme so that each new origin (before being randomized) is equidistant from two neighbouring centres that have already been computed.

Devise a similar system for a *close-packed* network of circles based on a network of three sets of equally spaced lines at 120° to one another, thereby forming a triangular network to define the origins of the perfect array.

15.3 A glass fibre 1 mm in diameter has an ultimate tensile stress of 40 MN m^{-2}. If the Young's modulus of glass is 80 GN m^{-2} estimate a lower limit for the surface energy of glass. What is the value of this energy per surface atom in eV?

15.4 The electron band gaps of a crystalline conductor occur at the boundaries of the Brillouin zones. On this basis show that in monovalent amorphous metals the value of k_F (i.e. the wave vector corresponding to E_F) should be a little less than $q_m/2$, where q_m is the wave vector (i.e. the abscissa in Fig. 15.8a) corresponding to the first maximum in the X-ray intensity distribution. Show that in divalent materials k_F should be slightly greater than q_m.

15.5 If the minimum electrical conductivity occurs when $kl = 1$, show that this conductivity is of the order of $e^2/(3\hbar a)$. (Use section 7.3, p. 108 and eqn (8.2), p. 125.)

15.6 From the high-temperature part of the curve of Fig. 15.13 estimate the value of σ_0 in eqn (15.16) and compare this value with the minimum metallic conductivity from problem 15.5.

15.7 Show that if variable range hopping (eqn (15.27)) occurs in a two-dimensional structure then the conductivity, σ, should have a temperature dependence $\ln \sigma \propto -T^{1/3}$.

15.8 Using the data of Fig. 15.12 estimate the value of B in eqn (15.28). If the electron wave function falls to zero over a distance of three atomic spacings, estimate the density of electron states at E_F. Discuss whether this seems to be a reasonable value.

15.9 In section 15.27 we made the assumption that the density of two-level-states, G_{TLS}, was constant. Measurements of the specific heat of a particular sample at very low temperatures show a temperature dependence of $T^{0.9}$. What is the energy dependence of G_{TLS}?

15.10 As discussed in section 15.28, the thermal conductivity of amorphous materials varies approximately as T^2 at very low temperatures. In fact experiments show that in many cases the temperature dependence is closer to $T^{1.8}$. What would the energy dependence of G_{TLS} need to be in order to account for this?

15.11 The specific heat at 5 and 10 K for the material whose thermal conductivity is shown in Fig. 15.17 is 4375 and 19000 J K^{-1} m^{-3}, respectively. If the sound velocity is 3000 m s^{-1} (approx. independent of T), calculate the effective phonon mean free path at 5 and at 10 K and discuss the significance of the values which you obtain.

15.12 Describe those characteristics of amorphous magnetic materials that make them suitable for technological applications. Discuss why they can be used to advantage in certain devices.

15.13 Draw an 8×8 square network similar to Fig. 15.19. Programme a random number generator to give 30, 40, 50, and 60 per cent of the values greater than unity. Go through the squares in order, filling them or leaving them blank, as the number is greater or less than unity respectively. Show that a percolating path across the array occurs at about 60 per cent occupancy. Plot the length of the path as a function of occupancy and see if you obtain a curve similar to Fig. 15.20.

If possible extend this procedure to an $n \times n$ array with a computer programme. Show that the rise at p_c becomes sharper as n increases.

Answers to problems

Chapter 1, pp. 16–17

1.1. 0·287 nm ; 0·248 nm.
1.2. 0·235 nm ; 0·144 nm.
1.3. 2345 kg m^{-3}.
1.4. 144·7° ; 0·173 nm ; 0·212 nm.
1.6. 0·304 nm.
1.7. 1·058 × 10^{-105} ; 2·520 × 10^{-28} ; 2·57 × 10^{-8} N ; 1·285 × 10^{-9} N.
1.8. 1·99 × 10^{-19} J.

Chapter 2, p. 35

2.1. 19·5° ; 41·8° ; 1 : 1.
2.2. (a) first order is zero. (b) 1 : 9.
2.3. 0·416 nm ; none, n.b. (100) and (110) are forbidden orders for f.c.c.
2.4. 4.
2.5. 3·12 × 10^{-21} J ; 151 K.
2.6. +7·5 ; −9·6 ; (a) neither, (b) both, (c) (030).
2.7. 3 to 18.
2.8. 1·91 × 10^{-5} K^{-1}.

Chapter 3, pp. 51–2

3.1. 1 eV.
3.2. (1 + 1·94 × 10^{-4}) : 1.
3.4. 1·77 × 10^6.
3.5. 4 planes, 6 directions.
3.7. 128 N.
3.9. About 10^{-9} J m^{-1}.
3.11. 114 h.

Chapter 4, pp. 76–77

4.1. 2 × 10^{-9} J m^{-1}.
4.2. 0·47 N m^{-1} ; 1·56 GPa.
4.3. 36 nm.
4.4. 29 MPa.
4.5. (a) 7 × 10^{-19} J ; (b) 0·7 N m^{-1} ; (c) 2·33 GPa.
4.6. 24 μm.
4.7. 6·7 × 10^{10} m^{-2}.
4.8. 1·75 × 10^{11} m^{-2}.
4.9. 15 nm.

Chapter 5, pp. 93–4

5.1. (a) ⩾ 144 K ; (b) 0·16 K ; (c) 0·136 K ; (d) $T = \infty$ required.
5.2. 6·3 × 10^{-97} J ; 6·3 × 10^{-8} J ; 71 J ; 2700 J ; > 500 K.
5.3. 7470 J mol^{-1} ; ∼4000 J mol^{-1}.
5.4. 169 K.
5.5. 1·55 × 10^{-2} J mol^{-1} ; 2·4 × 10^{-3} J mol^{-1}.
5.6. (a) ∼8000 J ; (b) ∼640 J ; (c) 2·89 J ; 3 l ; 0·24 l ; 10^{-3} l.

5.7. 1.2×10^4 m s^{-1}: 1 nm: 1.2×10^{13} Hz.
5.8 0.11.
5.9 (a) 0.42; (b) 0.91; (c) 1.44.

Chapter 6, p. 105

6.1. 30 nm.
6.2. 1953 : 1.
6.4. 403 : 1.
6.5. 1.9 W m^{-1} K^{-1}.
6.6. 3.3×10^{13} Hz: $\sqrt{2}\ aC/\pi$, $aC/2\sqrt{2}$: $-\sqrt{2aC}/\pi$, $-aC/2\sqrt{2}$: aC/π, 0: $\sqrt{2aC}/3\pi$, $-aC/2\sqrt{2}$: $-\pi/2a$ equiv. to $+3\pi/2a$ phonons.

Chapter 7, p. 123

7.1. 3.27×10^{-19} J; (a) 6.1×10^{40} J^{-1}; (b) 2.76×10^{42} J^{-1}: 2%.
7.2. 6.7×10^{-19} J: 1.2×10^6 m s^{-1}.
7.3. 0.15 J mol^{-1} K^{-1}; 3.3 K
7.4. 1.6×10^{-11} N^{-1} m^2.
7.5. 6.4×10^{-6}.
7.6. 10^{-4} V.
7.7. 2.
7.8 103, 207, 291 J (kg mole)$^{-1}$ K^{-1}.

Chapter 8, pp. 141–2

8.2. (a) 1.7×10^{-9} Ωm; (b) 3.7×10^{-9} Ωm.
8.3. 9.2×10^{-6} : 1.
8.4. $\pi/3$ (= 1.04).
8.6. $v_{phase} = \frac{1}{2} v_{group}$.
8.7. 13.5 nm; 1.16 m s^{-1}.
8.8 0.37 V.

Chapter 9, pp. 158–9

9.1. 3.4.
9.3. 4 K.
9.4. 8.4×10^{-4} Ωm.
9.6. 10^{-4} : 1.
9.7. 5×10^{-4} eV; 1700 $a_0 = 10^{-7}$ m.
9.8. 2 Ωm.
9.10. 5000 : 1.
9.11. 1.1×10^{-2} m_e

Chapter 10, pp. 185–6

10.1 $l_n = l_p = 0.81$ μm; 87 pF.
10.2 3.5×10^{25} m^{-3}.
10.3 5×10^8:1.
10.4 1.9 ms.
10.5 27 mA.
10.7 -5.15×10^{-20} J.
10.11 6×10^{-21}, 24×10^{-21} J; 9.4×10^{16} m^{-2}.
10.12 1.5×10^{17} m^{-2}.

Chapter 11, pp. 198–9

11.1. 1.25×10^{-7}.
11.2. 4.43×10^{-5}.
11.3. 3.4 T.
11.4. 5.6×10^{-24} J; 2.38; 0.55 T.
11.6. $1.18 : 1$.

Chapter 12, pp. 213–4

12.1. 1.7×10^{-24} J; 0.12 K.
12.2. 2100 T.
12.3. 28.7 nm; 1.15×10^{-3} J m^{-2}.
12.4. $7.7 \times 10^{-5} : 1$.
12.5. ~ 40 J m^{-3}; 2 W.

Chapter 13, pp. 231

13.1. 1.003.
13.2. 2.33×10^5 Vm^{-1}.
13.3. $12\pi\varepsilon\varepsilon_0 r^3 E_0/(2\varepsilon + 1)$.
13.4. 1.09×10^{14} Hz.
13.6. 35 GHz.
13.8. 1.9×10^{-6}.

Chapter 14, pp. 266–7

14.1. (a) 3.6×10^{-4} J; (b) 4.5×10^{-4} J K^{-1}.
14.2. 0.37 K.
14.4. $2.2 \times 10^4 : 1$.
14.5. (a) 53 nm; (b) 10^{28} m^{-3}.
14.7. 760 cm^3.
14.9. 70 mT; 0.1 μm; ~ -0.6.
14.10. 0.23 μA.
14.11. 8.6×10^{-15} W.

Chapter 15, pp. 300–1

15.1 7×10^{-4} m.
15.3 20 J m^{-2}; 7.8 eV.
15.6 $\sim 4.8 \times 10^5$ Ω^{-1} m^{-1}; $\sim 4.3 \times 10^5$ Ω^{-1} m^{-1}.
15.8 $\sim 6.3 \times 10^{46}$ J^{-1} m^{-3} ($= 9.6 \times 10^{21}$ (eV)$^{-1}$ cm^{-3}).
15.9 $G \propto E^{-0.1}$.
15.10 $G \propto E^{0.2}$.
15.11 20 nm; 4.7 nm.

Further reading

GENERAL REFERENCES

The following standard texts will give useful extra material for nearly all the topics covered in this book.

ASHCROFT, N. W. and MERMIN, N. D. (1976) *Solid state physics*, Holt, Rinehart, and Winston, New York.
A detailed, higher-level, modern treatment.

BLAKEMORE, J. S. (1985) *Solid state physics* (2nd edn), Cambridge University Press, Cambridge.
A lively intelligent presentation.

DEKKER, A. J. (1960) *Solid state physics*, Macmillan, London.
A good introductory text.

ELLIOTT, R. J. and GIBSON, A. F. (1974) *An introduction to solid state physics and its applications*, Macmillan, London.
A more advanced text.

KITTEL, C. (1986) *Introduction to solid state physics* (6th edn), Wiley, New York.
A standard text, brim-full with information and data.

WERT, C. A. and THOMSON, R. M. (1970) *Physics of solids* (2nd edn), McGraw-Hill, New York.
This is at a lower level than the other texts. It is very well written and illustrated.

Chapter 1: Atoms in crystals

BROWN, P. J. and FORSYTH, J. D. (1973) *The crystal structure of solids*, Edward Arnold, London.
A short, handy text which discusses X-ray, electron, and neutron diffraction.

GLAZER, A. M. (1987) *The structure of crystals*, Adam Hilger, Bristol.
A very clearly written, short introductory monograph.

Chapter 2: Waves in crystals

BROWN, P. J. and FORSYTH, J. D. (1973) *The crystal structure of solids*, Arnold, London.

CHAMPENEY, D. C. (1973) *Fourier transforms and their physical applications*, Academic Press, London and New York.
A good introduction and explanation of transforms and their applications.

CHAMPENEY, D. C. (1985) *Fourier transforms in physics*, Adam Hilger, Bristol.
An excellent introduction at undergraduate level.

GLAZER, A. M. (1987) *The structure of crystals*.

HOLMES, K. C. and BLOW, D. M. (1966) *The use of X-ray diffraction in the study of protein and nucleic acid structure*, Interscience, New York.
A reprint of a review article which gives an introduction to molecular biology diffraction studies.

LIPSON, S. G. and LIPSON, H. (1981) *Optical physics* (2nd edn), Chapter 3, Cambridge University Press.
A clear refreshing optical text which has a good introduction to transforms.

Chapter 3: Defects and disorder in crystals

HENDERSON, B. (1972) *Defects in crystalline solids*, Arnold, London.
A short introductory text.

HULL, D. and BACON, D. J. (1984) *Introduction to dislocations* (3rd edn), Pergamon, Oxford.
A short elementary treatment of dislocations and mechanical properties.
VAN BUEREN, H. G. (1960) *Imperfections in crystals*, North-Holland.
A large well-written source book.

Chapter 4: Dislocations in crystals
COTTRELL, A. H. (1964) *The mechanical properties of matter*, Wiley, New York.
A well-written standard text.
HULL, D. and BACON, D. J. (1984) *Introduction to dislocations* (3rd edn), Pergamon, Oxford.
READ, Jr., W. T. (1953) *Dislocations in crystals*, McGraw-Hill, New York.
A very clear exposition of dislocation geometry and theory.

Chapter 5: The thermal vibrations of the crystal lattice
ROSENBERG, H. M. (1963) *Low temperature solid state physics*, Chapter 1, Clarendon Press, Oxford.
ZIMAN, J. M. (1972) *Principles of the theory of solids* (2nd edn), Cambridge University Press.
A good well-written introduction to all solid state theory.

Chapter 6: Phonons in non-metals; thermal conductivity
ROSENBERG, H. M. (1963) *Low temperature solid state physics*, Chapter 3, Clarendon Press, Oxford.
ZIMAN, J. M. (1972) *Principles of the theory of solids* (2nd edn), Cambridge University Press.

Chapter 7: Free electrons in crystals
GREIG, D. (1969) *Electrons in metals and semiconductors*, McGraw-Hill, London.
A short elementary treatment with little mathematics.
MÜLLER, E. W. (1960) *Advances in electronics and electron optics*, **13**, 83.
A review article of field-ion microscopy with many illustrations.
SOLYMAR, L. and WALSH, D. (1984) *Lectures on the electrical properties of materials*, (3rd edn), Clarendon Press, Oxford.
An amusing elementary treatment of electrons in solids.
ZIMAN, J. M. (1972) *Principles of the theory of solids* (2nd edn), Cambridge University Press.

Chapter 8: Electrical conductivity and the band theory
ALTMANN, S. L. (1970) *Band theory of metals*, Pergamon, Oxford.
A very clear, rigorous, yet readable introductory treatment of band theory.
DUGDALE, J. S. (1977) *The electrical properties of metals and alloys*, Edward Arnold, London.
A very clear treatment which uses the minimum of advanced mathematics.
GREIG, D. (1969) *Electrons in metals and semiconductors*, McGraw-Hill, London.
ROSENBERG, H. M. (1963) *Low temperature solid state physics*, Chapters 4 and 5, Clarendon Press, Oxford.
SIEGBAHN, K. *et al.*, (1967) *ESCA*, Royal Society of Sciences Uppsala Ser. IV, vol. 20.
A useful source book on ESCA techniques and some early results.
SOLYMAR, L. and WALSH, D. (1984) *Lectures on the electrical properties of materials* (3rd edn), Clarendon Press, Oxford.
ZIMAN, J. M. (1972) *Principles of the theory of solids* (2nd edn), Cambridge University Press.

Chapter 9: Semiconductors

ADLER, R. B., SMITH, A. C., and LONGINI, R. L. (1964) *Introduction to semiconductor physics*, Wiley, New York.

MORANT, J. D. (1970) *Introduction to semiconductor devices* (2nd edn), Harrap, London.
A very good introduction to semiconductors and their uses.

PUTLEY, E. H. (1960) *The Hall effect and related phenomena*, Butterworths, London.
A moderately advanced text.

ROSENBERG, H. M. (1963) *Low temperature solid state physics*, Chapter 7, Clarendon Press, Oxford.

SOLYMAR, L. and WALSH, D. (1984) *Lectures on the electrical properties of materials* (3rd edn), Clarendon Press, Oxford.

VAN DER ZIEL, A. (1976) *Solid state physical electronics* (3rd edn), Prentice-Hall, Englewood Cliffs, NJ.
A large introductory text which covers a wide field and is often worth consulting.

Chapter 10: The physics of the semiconductor p–n junction

FRASER, D. A. (1986) *The physics of semiconductor devices* (4th edn), Clarendon Press, Oxford.

MULLARD, (1972) *Field effect transistors*, Mullard Limited Publications, London.
A simple technical publication.

MORANT, J. D. (1970) *Introduction to semiconductor devices* (2nd edn), Harrap, London.

NICHOLS, K. G. and VERNON, E. V. (1966) *Transistor physics*, Chapman and Hall (Science Paperbacks), London.
A useful introduction to semiconductor physics and to transistors.

SOLYMAR, L. and WALSH, D. (1984) *Lectures on the electrical properties of materials* (3rd edn), Clarendon Press, Oxford.

SPROULL, R. L. and PHILLIPS, W. A. (1980) *Modern Physics* (3rd edn), Chapter 10, Wiley, New York.
Contains a very good introduction to semiconductor physics.

VAN DER ZIEL, A. (1976) *Solid state physical electronics* (3rd edn), Prentice-Hall, Englewood Cliffs, N.J.

Chapter 11: Paramagnetism

BLEANEY, B. I. and BLEANEY, B. (1976) *Electricity and Magnetism* (3rd edn), Chapter 14, Clarendon Press, Oxford.
The chapters on magnetic properties are especially well presented.

MORRISH, A. H. (1965) *The physical principles of magnetism*, Wiley, New York.
A good general text on magnetism.

SLICHTER, C. P. (1978) *Principles of magnetic resonance* (2nd edn), Chapters 1 and 2, Springer, Berlin.
The first two chapters are a clear introduction to the subject. The later chapters are rather advanced and are more geared to workers in the field.

Chapter 12: Ferromagnetism, antiferromagnetism and ferrimagnetism

BLEANEY, B. I. and BLEANEY, B. (1976) *Electricity and magnetism* (3rd edn), Chapters 15 and 16, Clarendon Press, Oxford.

MORRISH, A. H. (1965) *The physical principles of magnetism*, Wiley, New York.

TEBBLE, R. S. (1969) *Magnetic domains*, Methuen, London.
A small monograph.

Chapter 13: Dielectric properties

ANDERSON, J. C. (1964) *Dielectrics*, Chapman and Hall, London.
A small introductory book.
BLEANEY, B. I. and BLEANEY, B. (1976) *Electricity and magnetism* (3rd edn), Chapter 10, Clarendon Press, Oxford.
PURCELL, E. M. (1985) *Electricity and magnetism* (2nd edn), Chapter 10 (Berkeley Physics Course, vol. 2) McGraw-Hill, New York.
A very clear explanation of fields in dielectrics.

Chapter 14: Superconductivity

KUPER, C. G. (1968) *An introduction to the theory of superconductivity*, Clarendon Press, Oxford.
LYNTON, E. A. (1969) *Superconductivity* (3rd edn), Chapman and Hall (Science Paperbacks), London.
MCCLINTOCK, P. V. E., MEREDITH, D. J. and WIGMORE, J. K. (1984) *Matter at low temperatures*, Chapters 4 and 8, Blackie, London.
ROSE-INNES, A. C. and RHODERICK, E. H. (1978) *Introduction to superconductivity* (2nd edn), Pergamon Press, Oxford.
A very clear introductory exposition.
ROSENBERG, H. M. (1963) *Low temperature solid state physics*, Chapter 6, Clarendon Press, Oxford.
SHOENBERG, D. (1952) *Superconductivity* (2nd edn), Cambridge University Press.
A clear description of the theoretical and experimental state of the subject before the B.C.S. theory.
TILLEY, D. R. and TILLEY, J. (1986) *Superfluidity and superconductivity* (2nd edn), Adam Hilger, Bristol.
A very good modern text which exploits the similarities of the two topics in its theoretical treatment.
TINKHAM, M. (1975) *Introduction to superconductivity*, McGraw-Hill, New York.
A good introduction with a strong theoretical bias.

Chapter 15: Amorphous materials

ELLIOTT, S. R. (1984) *Physics of amorphous materials*, Longman, London.
A very useful book which gives a general introduction to many topics, followed by quite extensive reviews of recent work.
HALL, C. (1981) *Polymer materials*, Macmillan, London.
A good simple introduction to the physical properties of polymers.
MARCH, N. H., STREET, R. A., and TOSI, M. (eds) (1985) *Amorphous solids and the liquid state*, Chapters 12–15, Plenum, New York.
A very useful source book.
MOTT, N. F. and DAVIS, E. A. (1979) *Electronic processes in non-crystalline solids*, Clarendon Press, Oxford.
The classic detailed text on electronic behaviour.
PHILLIPS, W. A. (1981) *Amorphous solids—low temperature properties*, Springer, Berlin.
A useful introductory research monograph.
STAUFFER, D. (1985) *Introduction to percolation theory*, Taylor and Francis, London.
An introduction which *really* is an introduction and is often entertaining.
ZALLEN, R. (1983) *The physics of amorphous solids*, Wiley, New York.
A good introductory text.

Index

Physical constants and conversion factors

Avogrado constant	L or N_A	$6 \cdot 022 \times 10^{23} \, \text{mol}^{-1}$

Bohr magneton	β or μ_B	$9 \cdot 274 \times 10^{-24} \, \text{J T}^{-1}$
Bohr radius	a_0	$5 \cdot 292 \times 10^{-11} \, \text{m}$
Boltzmann constant	k	$1 \cdot 381 \times 10^{-23} \, \text{J K}^{-1}$

charge of an electron	e	$-1 \cdot 602 \times 10^{-19} \, \text{C}$
Compton wavelength of electron	$\lambda_c = h/m_e c = 2 \cdot 426 \times 10^{-12} \, \text{m}$	
Faraday constant	F	$9 \cdot 649 \times 10^4 \, \text{C mol}^{-1}$
fine structure constant	$\alpha = \mu_0 e^2 c / 2h = 7 \cdot 297 \times 10^{-3} \, (\alpha^{-1} = 137 \cdot 0)$	

gas constant	R	$8 \cdot 314 \, \text{J K}^{-1} \text{mol}^{-1}$
gravitational constant	G	$6 \cdot 673 \times 10^{-11} \, \text{N m}^2 \, \text{kg}^{-2}$
nuclear magneton	μ_N	$5 \cdot 051 \times 10^{-27} \, \text{J T}^{-1}$
permeability of a vacuum	μ_0	$4\pi \times 10^{-7} \, \text{H m}^{-1}$ exactly

permittivity of a vacuum	ε_0	$8 \cdot 854 \times 10^{-12} \, \text{F m}^{-1} \, (1/4\pi\varepsilon_0 = 8 \cdot 988 \times 10^9 \, \text{m F}^{-1})$
Planck constant	h	$6 \cdot 626 \times 10^{-34} \, \text{J s}$
(Planck constant)/2π	$\hbar$	$1 \cdot 055 \times 10^{-34} \, \text{J s} = 6 \cdot 582 \times 10^{-16} \, \text{eV s}$

rest mass of electron	m_e	$9 \cdot 11 \times 10^{-31} \, \text{kg} = 0 \cdot 511 \, \text{MeV}/c^2$
rest mass of proton	m_p	$1 \cdot 673 \times 10^{-27} \, \text{kg} = 938 \cdot 3 \, \text{MeV}/c^2$
Rydberg constant	$R_\infty = \mu_0^2 m_e e^4 c^3 / 8h^3 = 1 \cdot 097 \times 10^7 \, \text{m}^{-1}$	
speed of light in a vacuum	c	$2 \cdot 998 \times 10^8 \, \text{m s}^{-1}$

Stefan–Boltzmann constant	$\sigma = 2\pi^5 k^4 / 15 h^3 c^2 = 5 \cdot 670 \times 10^{-8} \, \text{W m}^{-2} \text{K}^{-4}$	
unified atomic mass unit (^{12}C)	u	$1 \cdot 661 \times 10^{-27} \, \text{kg} = 931 \cdot 5 \, \text{MeV}/c^2$
wavelength of a 1 eV photon		$1 \cdot 243 \times 10^{-6} \, \text{m}$

$1 \, \text{Å} = 10^{-10} \, \text{m}$; $\quad 1 \, \text{dyne} = 10^{-5} \, \text{N}$; $\quad 1 \, \text{gauss (G)} = 10^{-4} \, \text{tesla (T)}$;
$0°\text{C} = 273 \cdot 15 \, \text{K}$; $\quad 1 \, \text{curie (Ci)} = 3 \cdot 7 \times 10^{10} \, \text{s}^{-1}$; $\quad 1 \, \text{atm} = 10^5 \, \text{N m}^{-2}$;
$1 \, \text{J} = 10^7 \, \text{erg} = 6 \cdot 241 \times 10^{18} \, \text{eV}$; $\quad 1 \, \text{eV} = 1 \cdot 602 \times 10^{-19} \, \text{J}$; $1 \, \text{cal}_{\text{th}} = 4 \cdot 184 \, \text{J}$;
$\ln 10 = 2 \cdot 303$; $\quad \ln x = 2 \cdot 303 \lg x$; $\quad e = 2 \cdot 718$; $\quad \lg e = 0 \cdot 4343$; $\quad \pi = 3 \cdot 142$;
$1 \, \text{mol occupies } 2 \cdot 24 \times 10^{-2} \, \text{m}^3$ at S.T.P.; $\quad 1 \, \text{micron} \, (\mu\text{m}) = 10^{-6} \, \text{m}$.